SEVEN SUBLIMES

SEVEN SUBLIMES

DAVID E. NYE

THE MIT PRESS
CAMBRIDGE, MASSACHUSETTS
LONDON, ENGLAND

The MIT Press would like to thank the anonymous peer reviewers who provided comments on drafts of this book. The generous work of academic experts is essential for establishing the authority and quality of our publications. We acknowledge with gratitude the contributions of these otherwise uncredited readers.

This book was set in ITC Stone and Avenir by New Best-set Typesetters Ltd. Printed and bound in the United States of America.

Library of Congress Cataloging-in-Publication Data

Names: Nye, David E., 1946- author.
Title: Seven sublimes / David E. Nye.
Description: Cambridge, Massachusetts : The MIT Press, [2022] | Includes
    bibliographical references and index.
Identifiers: LCCN 2021033922 | ISBN 9780262046923
Subjects: LCSH: Technology—Social aspects—United States. | Technology—
    United States—Psychological aspects. | Technology—United States—History. |
    Sublime, The.
Classification: LCC T14.5 .N933 2022 | DDC 303.48/3—dc23
LC record available at https://lccn.loc.gov/2021033922

10 9 8 7 6 5 4 3 2 1

CONTENTS

PREFACE

The sublime is a widely shared emotion that all human beings, regardless of their race, gender, or nationality, are capable of experiencing, for they all are endowed with the same bodily senses. When W. E. B. Du Bois visited the Grand Canyon, the railroads forced him to travel in a segregated railway car because he was Black, but this did not prevent him from appreciating its grandeur. Du Bois declared unequivocally, "I believe that all men, black and brown and white, are brothers, varying through time and opportunity, in form and gift and feature, but differing in no essential particular, and alike in soul and the possibility of infinite development." His meditation on the enormous chasm concluded with these words: "It is not—it cannot be a mere, inert, unfeeling, brute fact—its grandeur is too serene—its beauty too divine! It is not red, and blue, and green, but, ah! the shadows and the shades of all the world, glad colorings touched with a hesitant spiritual delicacy."[1] Du Bois understood that the capacity to experience the sublime is universal.

This book focuses less on formal philosophy than on personal experiences, such as visiting a national park, skyscraper, disaster site, battleground, or virtual reality. To experience the sublime, it is not necessary to travel to famous locations. During travel restrictions due to the pandemic in 2020–2021, many people discovered solace and inspiration in local microadventures. They camped in nearby parks; they climbed trees;

they disrupted routines; they stared at the night sky; they took walks in unfamiliar places. They discovered that "much of the work of cultivating awe is in paying attention to the small details around us."[2] Recent quantitative psychological studies suggest that a majority of all people have experienced the sublime by the time they are twenty years old.[3] In such moments, "awe basically shuts down self-interest and self-representation and the nagging voice of the self."[4] Recent studies confirm what philosophers have long said: confronted with the sublime, people commonly feel a sense of humility. These experiences of awe reduce self-interest and increase social cohesion. The sublime is a powerful individual moment, but it also has cultural effects, helping to hold groups together. This is a useful starting point for an historical assessment of the sublime, considered not as a static category but as an evolving realm of experiences with at least seven distinct forms.

ACKNOWLEDGMENTS

In 1994 I thought that my *American Technological Sublime* (MIT Press, 1994) provided the epitaph for the technological sublime. I was wrong. Instead, for the last quarter century new objects have been considered sublime. This general interest by itself might have prompted me to write a book, but I was spurred on by three invitations. In the fall of 2017, the University of Lodz in Poland invited me to give the keynote address at a conference on the technological sublime. The following year, the journal *Azimuth: Philosophical Coordinates in the Modern and Contemporary Age* requested a contribution to a special issue, published in 2018 as "What Comes after the Technological Sublime?" Finally, the journal *ISIS* asked me to review Alan G. Gross's *The Scientific Sublime.*[1]

When I talked about writing a study of new forms of the sublime with Katie Helke, my editor at the MIT Press, she immediately encouraged me. Three institutions then helped make it possible. In 2017, the University of Minnesota appointed me Senior Research Fellow in the history of technology, provided a congenial workspace at the Charles Babbage Institute, and granted me access to library and archival resources. At the same time, the University of Southern Denmark allowed me to retain access to its library after retirement. In 2019, I worked out the structure of the project during a summer residency at the Netherlands Institute for Advanced

Study in Amsterdam. Three members of its staff were particularly helpful: Trinette Zecevic-Boulogne, Astrid Schulein, and Ruud Van Veen.

Because writing a book is largely a solitary affair, occasional discussions with colleagues were not just helpful but vital. Many offered advice or pointed me to useful materials. At Virginia Tech, Richard Hirsh listened to an account of the project in 2018 and took an interest in its development. At the University of Maryland, Thomas Zeller discussed his work on constructed sublime experiences on scenic highways in Germany and the United States. At the University of Minnesota, Douglas Lewis for half a century has patiently tried to improve my understanding of philosophy, while the late Alan Gross, whom I first met when the project was well along, suggested sources for chapters 4 and 5. At Minnesota's Charles Babbage Institute, Tom Misa and Jeffrey Yost suggested materials that improved chapter 6. My Danish colleagues Jørn Brøndal, Thomas Ærvold Bjerre, Niels Bjerre Poulsen, Anders Bo Rasmussen, and Anne Mørk drew my attention to historical examples. At the University of Munich, Klaus Benesch shared his thoughts on cyborgs and his expertise in literary and cultural history, while Temple University's Miles Orvell directed me toward photographic representations of the sublime. Richard Haw shared with me his wide knowledge of New York City and nineteenth-century technology. Years ago, Professors Jeff Bolster and Marty Melosi were senior Fulbright scholars in Odense, and they provided insights into environmental history that proved useful in writing chapter 8. None of these fine scholars, nor the three anonymous peer reviewers selected by MIT Press, are responsible for errors or misconceptions that may remain.

This book is gratefully dedicated to Helle Bugge Bertramsen Nye, who believed in it from the start and endured my talking about it until the end.

TANGIBLE SUBLIMES

1

NATURAL

What is it like to experience the natural sublime? Charles Dickens could answer that question based on travel experiences. Arriving at Niagara Falls, he saw first, "two, white, great clouds rising up slowly and majestically from the depths of the earth," and coming closer he heard "the mighty rush of water and felt the ground tremble underneath my feet." He climbed down the steep bank to the foot of the American falls, "deafened by the noise, half-blinded by the spray, and wet to the skin." Boarding a small ferry that crossed the river to the Canadian side, he felt "stunned, and unable to comprehend the vastness of the scene," until he stood on Table Rock and was enthralled by the cataract's "full might and majesty."[1] He remained at Niagara for ten days, viewing it from every angle and at different times of day. The experience engulfed his senses, and it filled him with a sense of peace, as "the strife and trouble of daily life receded from my view." Had Dickens stayed until winter, he might have seen the great falls transformed into a great wall of ice, with water falling behind it (figure 1.1).

Different sublime experiences arise from confrontations with impressive natural sites, large-scale technologies, disasters, warfare, intangible features of nature, digital technologies, and ecological complexity. However, the terminology used to discuss the different forms of the sublime has long been limited, with the result that quite different experiences

On the Ice Mountain, Niagara Falls, U. S. A.

1.1 Niagara Falls, Ice Bridge. Crowds gathered on the winter ice from the 1880s until 1912, when several deaths led to prohibition of the practice. Courtesy of New York Public Library, http://link.nypl.org/B7zLwJVXQxSL4m8BKB33uwb.

often are spoken of as though they were similar. But gazing from the top of a mountain at a vista is not the same thing as looking at a metropolis from the observation deck of a skyscraper. Living through a military bombardment is not like visiting Niagara Falls. Looking at images constructed from Hubble Space Telescope data is not the same as experiencing a powerful earthquake or visiting a battlefield. In different ways, each of these experiences can be sublime, but an expanded terminology is needed to distinguish among them. As Ronald Hepburn has observed of the natural sublime, "it is still seriously possible to look on a substantial set of recorded experiences of the sublime as having a phenomenological centre—approached but maybe never captured by aesthetic theorising in all its variety." The realization that the natural sublime remains elusive although it still has a phenomenological core also applies to other forms of the sublime. Hepburn characterizes the natural sublime as being an experience that "combined, or fused, dread at the overwhelming energies of nature and the vastness of space and time with a solemn delight or exhilaration."[2] By extension, other sublimes concern the energies of humankind in constructing powerful systems and large projects, in living through and often contributing inadvertently to disasters, in mechanized warfare, and in scientific discovery. Each of these may inspire a combination of awe, dread, delight, and exhilaration.

Dickens had another sublime experience when, by the light of a full moon, he climbed Mount Vesuvius, the volcano outside Naples. Near the top, his party arrived at the "region of Fire, an exhausted crater formed of great masses of gigantic cinders, like blocks of stone from some tremendous waterfall, burnt up; from every chink and crevice of which, hot, sulphurous smoke is pouring out: while, from another conical shaped hill, the present crater, rising abruptly from this platform at the end, great sheets of fire are streaming forth: reddening the night with flame, blackening it with smoke, and spotting it with red hot stones and cinders, that fly up into the air like feathers, and fall down like lead." Dickens clambered over "the broken ground," smelled the suffocating sulfurous fumes, and feared "falling down through the crevices in the yawning ground." The party had to stop "every now and then, for somebody" missing in the smoke, while listening to the "hoarse roaring of the mountain." With all their senses aroused, they crossed "to the foot of the present Volcano," sat "down among the hot ashes at its foot," and looked up "in silence; faintly estimating the action that is going on within, from its being full a hundred feet higher, at this minute, than it was six weeks ago." Standing so close to the fiery mouth of the volcano, Dickers felt there was "something in the fire and roar, that generates an irresistible desire to get nearer to it. We cannot rest long, without starting off, two of us, on our hands and knees, accompanied by the head guide, to climb to the brim of the flaming crater, and try to look in." Amid the rumbling, on a trembling "thin crust of ground, that seems about to open underneath our feet and plunge us in the burning gulf below" they face "the flashing of the fire" and "the shower of red-hot ashes that is raining down, and the choking smoke and Sulphur." They become "giddy and irrational, like drunken men." Yet they press on and "climb up to the brim, and look down, for a moment, into the Hell of boiling fire below. Then, we all three come rolling down; blackened, and singed, and scorched, and hot, and giddy: and each with his dress alight in half a dozen places."[3] Dickens did not use the word "sublime" in his descriptions of Niagara and Vesuvius, but these were sublime encounters that assaulted all of his senses.

Such sublime experiences gradually became salient after 1700. The sublime first emerged during the Roman Empire, when the term was used to praise well-crafted texts, speeches, or outstanding architecture.[4] When the idea of the sublime was revived in the eighteenth century, it primarily referred to impressive natural sites, although some authors, including Edmund Burke, also described particular buildings as sublime.[5] He emphasized that the sublime experience begins with astonishment, "that state of the soul, in which all its motions are suspended, with some degree of horror." Like Dickens at Niagara Falls or Mount Vesuvius, the observer is struck dumb with amazement, and "the mind is so entirely filled with its object, that it cannot entertain any other, nor by consequence reason on that object." The sublime has an overwhelming grandeur, and it "hurries us on by an irresistible force."[6] It begins with powerful sensations that meld together in a rush of impressions, a sensory overload. For Burke, as Sandra Shapshay explains, "the pleasure of the sublime does not result from something like a chain of reasoning or free play of ideas," which would be closer to Immanuel Kant's view. Rather, in Burke "it results from a basic but unreflective cognitive appraisal" based on strong impressions.[7] Burke linked the sublime to sensations. As the philosopher Richard Shusterman has emphasized, "Burke's aesthetics is distinctively embodied, relying on an implicit naturalistic, empiricist ontology that affirms the close union of mind and body while claiming that mental contents are ultimately the product of sensations involving bodily effects."[8] Burke did not posit a mechanical cause-and-effect relationship in which sensations caused the sublime, however, for he recognized that the mind also shapes sensations. Yet regardless of whether one agrees with Burke or Kant, as Emily Brady summarizes in *The Sublime in Modern Philosophy*, the qualities that make an experience sublime can include "darkness, obscurity, greatness, massiveness, the tremendous, towering, dizzying, shapeless, formless, boundless, blasting, thundering, roaring, raging, disordered, dynamic, tumultuous,"[9] all terms that might be applied to Niagara Falls or to Mount Vesuvius.

Modern psychology recognizes not only the five senses that Aristotle defined (touch, taste, smell, hearing, and sight), but also four more. Three of these are located in the skin that, in addition to touch, can also sense temperature, pain, and the position of the body, or body awareness. In addition, three semicircular canals in the ears provide a sense of balance and the ability to feel acceleration and deceleration.[10] These additional senses make it possible to avoid burns and frostbite, to know without looking where the various parts of our body are at all times, and to maintain our balance when moving through space. Dickens needed these additional senses at Niagara Falls and Mount Vesuvius, both of which were disorienting and potentially dangerous. Likewise, an earthquake upsets the sense of balance, and the experience of flying would be greatly diminished without awareness of balance or acceleration. Multiple sensations are part of most sublime experiences, merging in a powerful overall impression. A motorcyclist explained to Rebecca Solnit "the infinitely subtle ways racers use their bodies to turn at high speeds and the incredible pleasure of those acts."[11] These maneuvers rely on balance and body awareness, and a feeling for sudden acceleration and braking, as well as the rush of the wind, the roar of the motorcycle, and the blur of visual impressions when speeding through a landscape. Something similar would be true of skateboarders, divers, or gymnasts. All nine senses function in daily life, and a sublime experience arouses them to a high pitch.

Kant focused on the mind more than Burke, emphasizing the thoughts after a sublime encounter.[12] He made the invaluable observation that the natural sublime falls into two broad categories: the "mathematical sublime," an encounter with extreme magnitude such as the Grand Canyon or a view from the top of a mountain; and the "dynamic sublime," an encounter with irresistible force, such as a hurricane or volcanic eruption. All forms of the sublime have these variant modes, which can be broadly defined as either encounters with landscapes (magnitude) or encounters with spectacles (engulfing forces). When experiencing the natural sublime, the observer at first feels overwhelmed, diminished, and insignificant. Climbing a mountain is a good example of an encounter with immensity, which as Solnit points out is often misunderstood to be simply the conquest of the landscape; "but as you get higher, the world gets bigger, and you feel smaller in proportion to it, overwhelmed and liberated by how much space is around you, how much room to wander, how much unknown." Out of necessity, during a climb attention is mostly fixed on the steep and often uneven trail, with occasional pauses to see the enlarging view. But at the summit, the view opens in every direction, and "the world doubles in size."[13] Kant noted that in such a sublime experience initial awe leads to a "momentary checking of the vital forces" followed by a recuperation and an abiding sense of wonder. The sublime compels an observer to realize inner mental powers that reach beyond the evidence of the senses, which have been overwhelmed.[14] Because the mind is able to conceive patterns and meanings that surpass the senses, the observer achieves an enlarged feeling of self-worth. For these experiences to occur, however, it is vital that one be in relative safety. Kant explained that in the face of tremendous forces, "we are all the more attracted by their aspect the more fearful they are, when we are in a state of security."[15] Dickens, for example, was best able to take in Niagara Falls from the security of an island, not from the unstable boat that took him there. At Mount Vesuvius, he relied on a guide to lead him to a safe viewpoint.

The sublime occupies a central place in Kant's philosophy, which he developed as religious beliefs weakened and gave way to the Enlightenment. Rather than anchor morality in holy scripture, he concluded that it had to arise from some form of universally available experience. The natural sublime emerged in Kant's thinking as that key experience; a feeling for the natural sublime made it possible to intuit the power of our rational faculties. In short, for Kant there was a connection between sublime experience and ethics. Sublimity roused the mind to reach for higher thoughts, and it made the observer aware that the mind could reach beyond the palpable world of the senses. As John Goldthwait summarized, in Kant "the sublime makes man conscious of his destination, that is, his moral worth. For the feeling of the sublime is really the feeling of our own inner powers, which can outreach in thought the external objects that overwhelm our senses."[16]

The technological sublime superficially resembles Kant's natural sublime. The observer again is amazed when confronted by a vast scene, such as the view from the top of a skyscraper, or a powerful force, such as the launch of a rocket carrying astronauts to the moon.[17] But there are important differences between these encounters and the natural sublime. Substituting a technological object, such as a railroad, for a natural object, such as a powerful waterfall, places humanity in a different position. It may seem that the only difference is in the object, not in the mind of the observer, but the meaning of the sublime changes decisively when based on a cultural construction. A natural object is an expression of external powers, while a technological object is a human product. When nature is the powerful source of experience, humankind is reduced to insignificance. But when awe is induced by human constructions, the experience may be identified with the conquest of nature, for example by conquering gravity in flight or containing a powerful river behind a dam. Humanity, often personified by the architect or the engineer, is exalted. The observer is not made aware of inner powers that reach beyond the visible world, but rather focuses on the human power to transform the visible world.

The intangible sublime offers another perspective. A new apparatus can make possible new perceptions, as Galileo realized when, through a telescope, he saw new features of the universe such as the mountains of the moon. The heavens have always been a source of awe. But during the Renaissance scientists discovered that Earth is not the center of the universe, and then found that the sun was but a minor star in a vast space. To many, this dislocation was a diminishment, marginalizing humanity. Kant understood that human beings existed in a tiny part of the cosmos, but he found this fact potentially sublime. As he put it, "If the grandeur of a planetary world in which the earth, as a grain of sand, is scarcely perceived, fills the understanding with wonder; with what astonishment are we transported when we behold the infinite multitude of worlds and systems which fill the extension of the Milky Way!" Such contemplations exemplified Kant's mathematical sublime, or the encounter with that which is absolutely great. Kant understood that the universe is "an abyss of a real immensity; in presence of which all capability of human conception sinks exhausted, although it is supported by the aid of the science of number."[18] Kant saw that for science to advance perception was not enough. An observer needed mediating instruments and mathematics, products of what he termed "Reason," to gain a larger grasp of the universe.

The examples that Kant used in his *Universal History and Theory of the Heavens* were not available to philosophers before astronomers demonstrated that the sun, not Earth, was the center of the solar system, and discovered that the sun was but a small star in a vast universe. Even for Kant, the sublime was not an absolute but a contingent category, based on perceptions shaped by new scientific knowledge. Were Kant alive today, he presumably would be enthralled by new technologies such as the Hubble telescope that have deepened our understanding of the universe. Kant already knew that the sublime features of the universe cannot be grasped by the senses alone. By combining Newtonian mathematics with astronomical observations, Kant rather accurately predicted the shape of the universe. Likewise, today, only through reasoning and the use of instruments that extend the senses can one begin to understand black holes or dark matter.

Burke and Kant defined the sublimes known to the eighteenth century; new technologies subsequently made possible additional forms. Kant identified the dynamic and the mathematical modes of the sublime that can be experienced in the natural world. This study presents other forms of the sublime that have a similar duality. This is not merely a matter of taxonomy. The accelerated movement of the early railroad offered an experience akin to Kant's dynamic sublime, but without moral absolutes or the supposition that the deity was involved in its meaning. The Olympian impressions when seeing the world from the top of a skyscraper or from an airplane offer an experience akin to Kant's mathematical sublime, but they are not the same. The present work defines and briefly explains six additional forms: the technological sublime, the disastrous sublime, the martial sublime, the intangible sublime, the virtual sublime, and the environmental sublime.

Why these seven sublimes? This project began by surveying books that dealt with the sublime published after 1990, to update what I had learned when writing *American Technological Sublime*. I next examined Google Scholar's list of more than two thousand articles that cited that book,[19] as well as articles listed in J-STOR, MUSE, and other bibliographic aids. This survey was not a matter of quantification. An article that merely mentions the sublime in a footnote did not merit the same attention as a seminal book such as Brady's *The Sublime in Modern Philosophy*. Likewise, an author who uses a term such as "postmodern sublime" or "digital sublime" with little attempt at a definition was less helpful than J. Glenn Gray's *The Warriors*.[20] However, casting a wide net established the range of objects that might be considered sublime and suggested topics that I might have overlooked, such as holography and drone photography. This survey included studies not only of impressive natural sights and technologies such as skyscrapers and railroads, but also of photographs, films, literary works, paintings, factories, canals, dams, battles, scientific discoveries, computer games, virtual reality, world's fairs, spectacular lighting, earthquakes, fires, floods, and more. I tried various ways to organize this material. A chronological approach did not suggest distinct eras that could be organized into coherent clusters for analysis. Divisions based on nation-states created tedious repetitions, as one would then need consider each topic—from volcanoes to canals to battles to virtual reality—in every national context. Instead, I saw the importance of the dividing line between sublime phenomena experienced directly through the senses and intangible sublime that can only be known through instruments, such as telescopes, microscopes, sensors, and computers. By focusing on this distinction, I could sort examples of the sublime into four forms of the tangible sublime (natural, technological, disastrous, and martial) and three forms of the intangible sublime (scientific, digital, and environmental). This book devotes a chapter to each of these. While developing these chapters, I realized that the different sublimes did not merely focus on different classes of objects. Each also implied a distinct perception of space and time and entailed a different teleology. The natural sublime is ultimately about human insignificance when confronted with the enormous age and apparently infinite space of the universe. In contrast, the technological sublime focuses on humanity's advance into the future. In the disastrous sublime the eruption of natural forces, as in an earthquake or a fire, compels attention to perils in the present. The martial sublime is also intensely focused on an immediate danger, but humanity unleashes its destructive forces. Study of the three intangible forms of the sublime yielded three additional perspectives. In short, the seven categories proved to be more than merely seven clusters of objects that human beings have considered sublime. Each of the seven sublimes presented a distinct perspective on space and time, which is to say that each one pointed toward a different understanding of humanity's place in the world. Future researchers may discern additional, but probably not fewer, sublimes. The evidence suggests there are at least these seven.

In searching for examples of the seven sublimes, I have not been concerned with whether a witness actually used the word "sublime," or was familiar with Burke's or Kant's philosophy. Someone facing battle or viewing images from the Hubble telescope may not employ the word "sublime," but their experience nevertheless may exemplify it. Whalers who were awed by the Canadian Arctic wrote home of its magnificent landscapes.[21] In 2019, many scientists were moved by the first images of a black hole, which brought tears to their eyes and provoked exclamations such as "We have seen what we thought was unseeable" and "This will leave an imprint on people's memories."[22] They had experienced the sublime, even if they did not use the philosophical term.

Just as Kant distinguished between two modes of the natural sublime (mathematical and dynamic), the other six forms of the seven sublimes also have two modes of experience: dynamic spectacles and vast landscapes. Spectacles sweep up the observer with irresistible force and mesmerizing movement. Landscapes overwhelm with their extent, bulk, or grandeur. The following chart lists the seven forms of the sublime in the order that they will be discussed.

One chapter is devoted to each of these seven sublimes. This introductory chapter has briefly discussed the natural sublime. Chapter 2 concerns the technological sublime, which has two modes: the landscapes made

Table 1.1 Seven forms of the sublime

Chapter	Form of sublime	Dynamic mode Spectacles	Mathematical mode Landscapes
1	Natural	e.g., Niagara Falls	e.g., Grand Canyon
2	Technological	balloons, railroads, airplanes, rockets	bridges, dams, skyscrapers, factories, cityscapes
3	Disastrous	conflagration, earthquake, flood	ruins, recreations
4	Martial	battle, bombing	panoramas, battlefields, airshows
5	Intangible	driving a Mars rover	microscope, telescope, etc.
6	Digital	virtual reality	server farms & computers
7	Environmental	symbiotic complexity of habitat	blighted ecologies; antilandscapes

possible by bridges, dams, skyscrapers and other constructions; and the dynamic experience of accelerated movement through space by balloon, canal, railroad, automobile, airplane, or rocket. Both modes are often seen as examples of technical reason. Chapter 3 examines the disastrous sublime, which can be classified as either landscapes seen after the fact, such as the ruins of Pompeii, or as spectacles directly observed, such as the Great Chicago Fire of 1871. Chapter 4 examines the martial sublime, whose two modes are the dynamic experience of battle, including explosions, firefights, and bombardment, and the panoramic view of a battle when observed from a distance, preserved at a historic site, or recreated in painted panoramas or films. Anyone, regardless of race, gender, class, or culture, has the capacity to directly experience these tangible sublimes.

The argument then turns to experiences that are only possible with the aid of mediating technologies. Chapter 5 discusses the intangible sublime or encounters with objects or forces that human beings cannot observe with their unaided senses, such as microscopic particles or black holes. These sublime experiences are delayed by a mediating technology such as an orbiting telescope that sends back to Earth data that must be processed before they can be interpreted. The public's experience of such sublime discoveries is in keeping with the ideas of Longinus, the ancient author who first wrote of the sublime. He defined the sublime as the effect produced by persuasive oratory or writing that uplifts and enthralls an audience. Some scientists make discoveries accessible to the layperson, acting as rhetoricians who explain the microscopic building blocks of life, subatomic particles, or distant galaxies. There is also a second mode of the intangible sublime that is a dynamic experience, such as the exploration of Mars or another inaccessible location through the interactive control of a remote vehicle.

Mediated interactions are also possible through computers, and chapter 6 turns to dynamic digital experiences such as virtual reality. Not every engrossing digital technology is sublime, however, and this chapter briefly considers what experiences might qualify. There is also a landscape mode of the digital sublime, in which one is enthralled by the vast scale of information, the immensity and power of computer technologies like the Internet, and the enormous capacity of giant server farms.

Chapter 7 examines the environmental sublime, which neither celebrates the domination of nature, as in the classic technological sublime, nor seeks to construct a virtual nature, as in the digital sublime. In its dynamic mode, it focuses on living ecosystems, evoking wonder at their complex symbiosis. In contrast, the landscape mode of the environmental sublime confronts the death of ecosystems, due to pollution, habitat destruction, global warming, and species extinction. Human actions threaten not just one or another species but entire ecological systems such as coral reefs and rainforests. The environmental sublime expresses awareness of the consequences of human action. One cannot go back to the natural sublime of Kant or Ralph Waldo Emerson. In the Anthropocene, human beings need to move beyond celebrating the technological sublime, beyond the pretense that they are not part of nature, and beyond the melancholy contemplation of apocalypse, into a creative engagement with the environment.

Chapter 8 examines how sublimes have been selectively integrated into different national identities. Because sublime experiences are universal, it might seem that they would unite humanity across national borders. However, different cultures value somewhat different experiences. Every country develops a distinctive repertoire of possible sublime experiences, and these are interpreted differently from one nation to the next. The final chapter reviews the formations of the sublime, both those known directly through the senses and those experienced indirectly through mediating technologies. It contrasts the largely static landscapes with the dynamic, often dangerous spectacles, followed by discussion of the temporal and spatial dimensions of the different sublimes. Finally, the epilogue ponders the future of the sublime.

The first seven chapters organize the hodgepodge of places, events, and experiences that are frequently referred to as sublime. Because each of these seven sublime forms entails a fundamentally different concept of space and time, they can be on odds with one another. A waterfall or canyon exemplifies the natural sublime to some, and yet other people value more highly a large hydroelectric dam that obliterates these landscapes and exemplifies the technological sublime. Likewise, virtual reality (VR) makes possible new perceptions but engages only a few of the senses, in contrast to the all-encompassing sensory engagement with a local ecology that is the hallmark of the environmental sublime. The martial sublime and the technological sublime are based on the mastery of many of the same technologies, but they work toward quite different ends and express incompatible values. In short, the seven sublimes share certain characteristics, but they are not a coherent system. They are related but not congruent.

And yet, most of the seven sublimes were suggested in the writings of Kant and Burke, although both writers focused their attention on the natural sublime. Burke considered some buildings to be sublime, anticipating the technological sublime, discussed in chapter 2.[23] Kant wrote of violent storms and Burke of raging floods, suggesting the disastrous sublime, discussed in chapter 3. After seeing an inundation in Dublin, Burke wrote to a friend, "It gives me great pleasure to see nature in these great though terrible scenes. It fills the mind with grand ideas."[24] Kant considered warfare potentially sublime, while Burke speculated on the psychological impact of hearing cannons, adumbrating the martial sublime that is the subject of chapter 4.[25] Kant was acutely aware of the usefulness of telescopes and other tools that extended the senses to the intangible, and he helped lay the groundwork for the scientific sublime discussed in chapter 5. In his *Universal Natural History and Theory of the Heavens* he speculated that the nebulae were island universes, that there were likely planets beyond the solar system, and that new worlds were continually being created.[26] In short, the seeds of the different sublimes discussed in the first five chapters are in Kant and Burke. Furthermore, the environmental sublime, discussed in chapter 7, can be seen as a hybrid of the natural and the scientific sublimes, which leaves only the digital sublime as a largely new formation, but one that often imitates the others.

With this overview in mind, we can now consider the two modes of the technological sublime.

2

TECHNOLOGICAL

When the Danish writer Hans Christian Andersen traveled by train for the first time in 1840, he was amazed by how smoothly it glided over the German countryside. He seemed to be flying like a bird through the landscape or moving like clouds driven before a storm. He thought it far superior to travel by coach and declared that Faust could not have moved more quickly when enveloped in the cloak of his guide, Mephistopheles. With the railroad, humankind had taken control of natural forces to become as strong as medieval people thought the Devil must be. Indeed, it seemed humanity would soon leave him in the dust. Andersen had seldom been gripped so hard by any experience, and he compared it to coming face to face with God. It had the spiritual intensity of a childhood visit to a cathedral or seeing the starry heavens on a clear night by the sea.[1] In short, Andersen had experienced the technological sublime, and he had done so like thousands of others, by riding on the railroad.

The technological sublime emerged during the late eighteenth century. As landscape, it reconfigured Kant's mathematical sublime to include impressive human creations such as panoramas, bridges, dams, canals, world's fairs, skyscrapers, spectacular lighting, and scenic highways. As spectacle, it resembles Kant's dynamic sublime and concerns rapid or gravity-defying movement, including hot air balloons, railroads, steamships, zeppelins, automobiles, airplanes, and rockets. In some cases, notably the vistas of vast factories, canal zones, open-pit mines, and other immense projects, a single site is both a vast landscape and a dynamic spectacle. This chapter discusses a range of international examples.[2]

The constructed landscapes of the technological sublime suggest humanity's triumph over space. The mountain-top view presents the mathematical sublime of nature, but the vista from the observation deck of a tall building or along a waterfront reveals a vast, humanly constructed scene that replaces nature. The streets, bridges, and varied architecture provide the overall impression of society guiding powerful forces and actors. The public's fascination with such views was evident in the success of the first panoramic painting by Robert Barker and his son Henry. It gave a 360-degree view of Edinburgh in 1789. Their panorama of London in 1792 was even more successful with the public. Initially, it was a half-circle of 1,479 square feet that presented a view of the Thames River, its three bridges, and the cities of London and Westminster. The Barkers' advertisement for their panorama assured patrons that it "appears as large and in every respect the same as reality," a claim that became more accurate when the painting was extended to a full circular view the following year.[3] Many early panoramas depicted large cities and, like the one in London, they were a popular success, including views of places few could then afford to visit in person, such as Constantinople, Venice, Rome, and Athens.[4] The Barkers erected a rotunda in Leicester Square that displayed two panoramas, and had a workshop preparing new ones. One reviewer in 1803 noted that "most panoramas have made cities, ships or battles—that is to say the work of human hands—the center of attention."[5] Visitors climbed into an elevated viewing stand at the center of a round building and looked at a famous site painted on an encircling canvas. Props laid out from the foreground to the walls seamlessly blended in with the painting, enhancing the illusion of seeing an actual city. Such rotundas were built in France, Germany, Belgium, The Netherlands, and the United States. In 1799, Parisians flocked to a panorama of Paris and to another depicting a victory over the British and Spanish at Toulon.[6] Americans enjoyed imported European panoramas but also painted their own, notably one of Niagara Falls that placed the viewer "upon the beetling cliff of the mighty chasm," able to survey a "vivid and giddy prospect," a "sublime scene—the surrounding country, rivers, precipices, islands—the falling of the mighty torrent." Americans also developed a second form of panorama, an extremely long canvas that was unrolled across a stage, usually accompanied by music and narration. One of these, now lost, was advertised as "the largest panoramic painting of the scriptures in the world, being two-thirds of a mile long." Another canvas that has partially survived is 350 feet long and depicts a voyage down the Mississippi River.[7] In either format, the panorama provided an Olympian perspective that adumbrated the later enthusiasm for observation decks on skyscrapers.

Even as the panorama aroused popular enthusiasm by exemplifying the conquest of space, in France the Montgolfier brothers introduced the spectacle of hot-air ballooning that conquered gravity. Raising a balloon into the heavens was regarded as a stupendous achievement, and in 1783 half the population of Paris, about 400,000 people, saw one of the first ascensions.[8] They came from all social classes and, as is often the case when viewing a dramatic spectacle, they temporarily bonded in their enthusiasm. Today, a balloon seems to move slowly, but in the eighteenth century traveling twenty miles an hour with the wind was the outer limit of human mobility, and balloons effortlessly could move that fast. Moreover, rising into the sky seemed magically to open the heavens to conquest. Marie Thébaud-Sorger observed that "the sublime emerges from the beginning of the flights; the sublime as well as technological progress go hand in hand."[9] Benjamin Franklin, who then was the American ambassador to France, watched the first launch from the comfort of a carriage. He reported to a friend, "It diminished in Apparent Magnitude as it rose" until "it enter'd the Clouds, when it seemed to me scarce bigger than an orange."[10] Ballooning became a public craze, as crowds turned out to see it. In 1819 a diarist in New Orleans noted, "A great flock of people going to see the balloon rise; interesting group; men with their wives; *gentlemen* who never had any; negroes of all shades and sizes; dandies, sailors, boatmen, ladies," a crowd that increased for an hour, and then "the balloon ascends, a shout of triumph, sublime scene!"[11] Responding to such public enthusiasm, some entrepreneurs made a living at fairgrounds and on festival days, where paying customers could ascend in a tethered balloon, watched by an amazed crowd. Like Franklin, the people on the ground were awed to see the balloon shrink as it ascended, spellbound by its drifting movements, and deeply concerned for the safety of the passengers.

The few who rode in a balloon had a more intense experience. Drifting in the windy silence above the earth was both frightening and exhilarating. As Caren Kaplan notes, "flight made possible unprecedented aerial vistas," and demanded a retraining of the eye to understand the landscapes below.[12] Early balloonists felt exalted by the new perspectives and enraptured by sailing through the clouds. This "serene joy"[13] was part of a kaleidoscope of shifting experiences. Balloonists had to operate an unfamiliar mechanism while enduring discomfort and facing mortal danger. They might suffer cold, rain, or earache caused by the change in altitude, and they could crash or be blown out over water and drown. They conducted scientific experiments with barometers and thermometers and sketched the strangely flattened landscape below them. They were amazed by the sensations of buoyancy and the stillness of the ride. They experienced, in rapid succession, stunning vistas, windy silence, disorientation, fear, and the riveted attention of the crowds below.

Engineering projects that were celebrated as sublime included dams, bridges, factories, and skyscrapers.[14] The skyscraper emerged in the United States in the 1870s. The term came from the name of the highest sail on a ship, and John Moser used it to describe tall buildings in 1883 in "American Architectural Form of the Future." He declared, "This form of skyscraper gives that peculiar refined, independent, self-contained, daring, bold, heaven-reaching, erratic, piratic, Quixotic, American thought."[15] Replacing masonry construction with steel made the early skyscrapers of New York and Chicago feasible. They had elevators as well as stairs, electrical lights instead of gas, and telephones instead of employees to convey messages. The height of office towers and apartment buildings was determined by balancing the rental value of floor space against the rising costs of construction for higher floors. The tallest buildings could be further justified by the public relations value of owning a landmark. When constructed, the forty-seven-story Singer Building (1908) and the fifty-five-story Woolworth Building (1913) were the tallest in world. Their prominence generated valuable publicity for the Singer sewing machine and the Woolworth dime-store emporiums.

Adnan Morshed's *Impossible Heights* deals with the modernist city of skyscrapers as drawn by Hugh Ferris, imagined by Buckminster Fuller in his "aesthetics of ascension," and modeled by Norman Bel Geddes at the New York World's fair of 1939. All three saw themselves as master builders, and flight inspired them.[16] To create a simulated airplane ride at General Motors' Futurama pavilion, the fair's most popular exhibit, Bel Geddes commissioned aerial photographs. He then built a scale model based on careful analysis of how different environments appeared from a low-flying plane, to create "the impression of a continental flight over various types of terrain and urban regions."[17] Futurama gave visitors the magisterial vision of a sublime technological landscape.

Much urban photography also celebrates such constructed landscapes. Davide Deriu has studied "rooftop photography," or images made from tall buildings, radio towers, and cranes. He argues that the "plunging perspectives enhanced the overwhelming presence of the skyscraper" both as a "visual subject and viewing platform." While he looks at examples from several cultures, he argues that "the skyscraper epitomized a quintessentially American brand of modernity" creating the possibility of an Olympian gaze at the constructed urban landscape below. "By seizing possession of that dominant gaze, photographers" such as Berenice Abbott "revealed the contrast between the rampant high-rise constructions and the historical buildings and streetscapes that were dwarfed by them. This tension between old and new is rendered with dizzying intensity in a photograph such as her 'Broadway and Rector from Above.'"[18] Likewise, the writings of William Dean Howells, Theodore Dreiser, Frank Norris, and Jack London developed a sublime rhetoric and constructed "cognitive maps" of the otherwise dispersed and fragmented metropolis.[19] The view from a skyscraper became the analog of such a cognitive map, at once concrete and abstract, and it retains its appeal as new skyscrapers press ever higher, many of them designed not as offices but as exclusive apartment buildings.[20]

Skyscrapers also adopted spectacular lighting based on a tradition that goes back to Renaissance Italy, when fetes and civic celebrations were highlighted by fireworks, torches, and arrays of lanterns.[21] Illuminations spread throughout Europe in the following centuries, at first largely restricted to royal weddings, inaugurations, and the like, but gradually spreading to the rest of society. Lighting intensified and added dazzling special effects, and it was frequently described as sublime. Gas lighting after 1810 and electric lighting after 1875 intensified illumination far beyond mere functionality and transformed urban space. Charles Brush, Thomas Edison, Joseph Swan, and other inventors staged spectacular demonstrations of their lighting systems. After circa 1880 these powerful visual experiences intensified in the extravagant lighting of world's fairs and city centers. The night landscape was continually expanded and reached an early apotheosis in the cities of the United States, especially New York.[22] Painters, poets, and guidebooks celebrated the brilliant illumination, and other cities aspired to have "Great White Ways" and to be known through illuminated landmarks like the Statue of Liberty, the Empire State Building, and Brooklyn Bridge. Like earlier forms of the technological sublime, however, electrical displays soon became familiar and ceased to seem sublime, which stimulated the development of still more powerful special effects. By the late twentieth century the epiphany experienced in the electrified city had become a rush of simulations.[23] Furthermore, as separate generating systems were linked into larger regional grids, wide-scale electrical blackouts became possible. These periodically defamiliarized the urban night.[24] The city became a heterotopian landscape, whose electrical displays and occasional blackouts destabilized the sense of space, culminating at sites such as world's fairs, amusement parks, Times Square, and Las Vegas. These presaged the online spectacles of the digital sublime.

Spectacular lighting spread to tourist destinations worldwide, including Notre Dame and the Arc de Triomphe in Paris, the Colosseum in Rome, and Piccadilly Circus and the Houses of Parliament in London. Lighting arrays were also installed at natural sites such as Old Faithful geyser at Yellowstone National Park and the Carlsbad Caverns. Emily Marlowe found that seeing the caverns' underground waterfall bathed in electric light was a sublime experience that impressed Americans who "integrated it into their perceptions of what nature and technology could include."[25] The dramatic use of electric lighting also transformed amusement parks. Lauren Rabinovitz notes that the "numerous postcard views of amusement parks across the country offered the technological sublime in perfect miniature." In contrast, "the movies constructed it as a gigantic panorama" on large screens. Both "postcards and movies staged in these views dreams of modernization" at a time of rapid cultural change.[26]

These dreams of modernization were actualized through increasingly rapid communication and transportation, whose accelerating speed measured the conquest of time. Canal boats moved at the pace of a walking horse; nineteenth-century railroads at first went 20 miles per hour and gradually increased to 100; twentieth-century airplanes soon doubled that. Faster communication likewise sped up the flow of information, and during the nineteenth century the telegraph and later the telephone seemed sublime wonders, demonstrating humanity's triumph over distance. Individuals also moved ever faster in a sequence that included not only bicycles and automobiles, but also downhill skiing, motorboats, and snowmobiles. As the cultural geographer J. B. Jackson observed, speed provides a "sense of danger or at least of uncertainty, producing a heightened alertness to surrounding conditions." None of these activities focused on the landscape. It became almost abstract as it was "seen at a rapid, sometimes even a terrifying pace."[27] People wanted a maximum of experience in a minimum of time.

The automobile facilitated more freedom of movement than the railroad, and automotive tourism became an accelerated form of the country ramble. Many travelers went off the main line in search of sublime natural sites, notably at national parks, which vigorously promoted automotive tourism after around 1918. Some roads were built not to improve travel between regions but rather to accommodate those searching for novel landscape experiences. A notable early example was the Columbia River Highway (1913–1922) "which established the state of the art for building scenic roads in mountainous areas." It had naturalistic tunnels with dramatic openings that framed river views, and rustic guardrails and other safety features that blended into the surrounding landscape. There also were dramatic excursions up through hairpin turns to high ground (figure 2.1).[28]

Several roads inside or near national parks became nationally famous, notably the Trail Ridge Road in Rocky Mountain National Park and the Going-to-the-Sun Highway in Glacier National Park.[29] These "followed high-elevation, ridge-top routes that no ordinary highway would ever

2.1 Columbia River Highway, hairpin turn near Troutdale, Oregon. Initial construction, 1913–1922. Courtesy of Library of Congress, Prints and Photographs Division, https:// tile.loc.gov/storage-services/master/pnp/habshaer/or/or0300/or0386/photos/354759 pu.tif.

traverse."[30] Spectacular driving became a conscious goal of National Park Service highway engineers. Their roads were not built to connect cities or to speed travel but rather, to provide dramatic experiences. Another early example is the Needles Highway built inside South Dakota's Custer State Park in the early 1920s.[31] It did far more than provide tourist access to rock formations; "it took them into the Needles, wound around the formations and passed through tunnels painstakingly blasted out." Drivers were thrust close to "towering granite outcroppings" that "contrasted dramatically with the distant vistas. The tunnels, the twisting road, and the numerous switchbacks and the sharp drop-offs provided drivers with an exhilarating experience."[32] One of the first to travel the highway lyrically described "looking out through the windows of Heaven, down upon the mountain peaks, so far below that the homes of men are as sparrow's havens . . . and still above is the great maze of the Cathedral Spires."[33]

In the following decade, when Mount Rushmore was being carved, South Dakota built the Iron Mountain Road that incorporated the giant sculpture of four U.S. presidents into its design. The motorist caught glimpses of the great stone faces at several points during the trip, until suddenly the end of a tunnel framed George Washington like a gigantic cameo. The road fused the sculpture and the landscape into a form of the technological sublime. During the 1930s the much longer Blue Ridge Parkway and several German roads built in the Alps also became destinations in themselves, providing not only the dramatic vistas traditionally associated with the mathematical sublime but also a dynamic experience of unfolding space revealed through sudden twists, hairpin turns, and tunnels.[34] As intended, such automotive landscapes attracted tourists, some driving auto campers and recreational vehicles, whom Peter Krapp has called the "Nomads of the Technical Sublime." These nomads, he wrote, appreciate roads as the epitome of "the technological sublime of contemporary landscapes." To capture such experiences cinematically, films often use high angle shots that do not match what an RV driver sees but rather depict sweeping landscapes while miniaturizing the vehicles.[35] In this way, the shots capture both the landscape and movement through it.

The Panama Canal also attracted tourists. It fused a constructed landscape with a dynamic spectacle. Its practical purpose was to help ocean

2.2 Gatun Locks, Panama Canal, c. 1915. Courtesy of Library of Congress, Prints and Photographs Division, https://www.loc.gov/item/2016820934/.

liners, freighters, and naval vessels move between the Atlantic and Pacific oceans without the time-consuming circumnavigation of South America. But it also became a popular example of the technological sublime. In *Seaway to the Future* Alexander Missal examines how the Canal Zone became a tourist site by 1910, when it was still being constructed.[36] Visitors marveled at the scale of the excavations, hydroelectric dams, powerful lights, the electric railway that towed the ships, and the enormous Gatun Locks (figure 2.2). Many considered the Canal Zone a tropical utopia, where humankind controlled and improved upon the natural world and built a model community for those operating the sealink. Yet the Panama Canal was a military enclave in the new country of Panama that had been torn from Colombia by a convenient coup d'état just before construction began. The thousands of manual workers brought to the site lived in crowded quarters, worked in tropical heat, and often contracted malaria. When the canal was completed, most of them were no longer needed. The Panama Canal strengthened the U.S. sphere of influence in the Western Hemisphere. It was a technological achievement that shortened sea voyages, improved trade, exemplified the subjugation of nature, and projected American power. It provided sublime views of a constructed landscape and demonstrated formidable technologies. But as with many large projects, including railroads, canals, factories, mines, and dams, the human cost of the technological sublime tended to disappear from popular memory.

In 2013, the Hudson River Museum held an exhibition about the industrial landscape along New York City's rivers between 1900 and 1940.[37] It brought together paintings by George Ault, Georgia O'Keefe, Oscar Bluemner, Daniel Putnam Brinley, and many more. The *New York Times* reviewer noted that these artists "were seeking transcendence in the great, gritty machinery of the city, just as previous generations of painters had sought it in the rough grandeur of the American wilderness." Their paintings depict powerful creations such as bridges, railroad yards, and skyscrapers, in which human beings themselves are minuscule and overwhelmed. The canvases depict "agony and ecstasy, shimmery smoke and hard-edge steel, grim murk and eye-popping color—the early 20th-century city was all these things."[38] The technological sublime also can be seen at enormous factories. Early examples include the enormous Krupp steelworks in Germany, or Henry Ford's River Rouge automobile plant that in 1930 was the world's largest factory.[39] Such sites combine the dynamic movement of production and a vast mechanized landscape. At the River Rouge plant the river was transformed into a canal, the land was flattened, and the placement and interior organization of every building was determined by the flow of the assembly line from raw materials on the docks to completed automobiles. Thousands toured such factories, which were consistently described in terms of the sublime.[40] Charles Sheeler made a celebrated series of photographs of the River Rouge plant, followed by several paintings of it that became synonymous with the technological sublime. As Leo Marx wrote of Sheeler's "American Landscape" completed in 1930, in it "no trace of untouched nature remains. Not a tree or a blade of grass is in view. The water is enclosed by man-made banks, and the sky is filling with smoke." In this orderly vision, "every natural object represents some aspect of the collective economic enterprise. Technological power overwhelms the solitary man."[41]

The imposition of an idealized order could also be seen on the H. J. Heinz factory tour in Pittsburgh. "Each stop focused on a sublime or dramatic aspect of the plant. The Time Office building, for example, was a miniature replica of the Library of Congress in Washington, D.C.; the Pickle Bottling Department offered a view of several hundred women in identical white caps and aprons stuffing pickles into glasses."[42] In the 1920s more than two hundred thousand people annually toured the Armour meatpacking plant in Chicago, and thousands more visited General Electric in Schenectady, the Hershey chocolate factory in eponymous Hershey, Pennsylvania, or Nabisco's Shredded Wheat factory at Niagara Falls. The popularity of industrial tours waned in the second half of the twentieth century, however, replaced by company visitor centers that featured films, dioramas, and shopping.[43] Usually the items for sale included photographs that depicted the factory as a landscape of production in the tradition of the sublime.

Another version of the technological sublime emerged in the landscapes of open-pit mining. The first open-pit mine was begun in 1906 outside Salt Lake City, at the head of Bingham Canyon. The Utah Copper Company mined low-grade ore with enormous steam shovels, and gradually, "girdled the mountain with a corkscrew of wide benches steadily tightening inward." In the next half-century, the mountain disappeared, as they removed half a billion tons of rock. By the 1930s the mine was spiraling downward into a growing pit, and by 1970 it had produced more copper than any other mine. Eventually, the company had removed six billion tons of ore, making it the largest excavation in the world (figure 2.3).[44] Generally, as high-grade deposits were mined out, open-pit mining of low-grade ore took its place. No longer tunneling underground, companies removed all the rock, ground it into sand, and then separated out the metal. In Butte, Montana, the Anaconda Company adopted this practice. In the 1950s its "brochures emphasized people's awe at the enormous size of the equipment including the 'mammoth trucks' that moved the material, the great cost of the operations, the abundance of ore that comes out of the pit and most importantly, the 'spectacular' expanses of the pit itself."[45]

2.3 Bingham Mine, Utah, August 1972. Jack E. Boucher, photographer. Courtesy of Library of Congress, Prints and Photographs Division, https://www.loc.gov/pictures/item/ut0029.photos.157721p/.

Such mining also seemed sublime at a Japanese pit in Manchuria. In 1928, it inspired the feminist poet and social activist Yosano Akiko when visiting the site, to write:

We were amazed by the magnificent sight of the huge open pit mine in the old city of Fushun. It immediately challenged the notion of a coal mine as an excavation of horizontal and vertical tunnels dug deep into the ground, for one need only peel off the outermost layer of oil shale—thirty or forty feet thick—to expose all of the coal beneath. I initially thought that this was the frightful and grotesque form of a monster from the earth, opening its large maw towards the sky. However, after going down a little and standing atop the manmade steps, I sensed a certain grandeur a few times greater than that of a large coliseum from Roman times, and felt that human beings who would use nature in such a way were like an intelligent species of ant.[46]

Today, only the most casual visitor to such a site can ignore the environmental costs of open-pit mining. Waste from the Bingham pit, for example, became one of the worst sources of toxic pollution in the United States, a problem that remains unresolved a century after mining began.[47]

Nevertheless, for more than a century, novelists, artists, photographers, and tourists visited mines, steel mills, and factory complexes and appreciated them as examples of the technological sublime.[48] The cultural geographer John Stilgoe observed that in the decades before World War I writers such as Arnold Bennett and Theodore Dreiser praised factory districts for their vast, smoky factories.[49] Adopting an impressionistic style, they presented the industrial world as a separate realm that was fascinating because it was utterly unnatural. Manufacturing districts seemed aesthetically pleasing from certain vantage points, along a canal, from the top of a skyscraper, or from a train window. Pittsburgh was often described as sublime, particularly at night. Historian Martin Aurand argues that the enormous Westinghouse Works in that city exemplified the technological sublime. The "sheer size and the repetition of elements" of this factory complex suggested infinity. The Westinghouse Works had "a continuous brick facade, displaying window after window after window. For the spectator on the surrounding hills, the complex was a vast expanse of roofs, nearly filling the valley from side to side." The huge factories at night were lighted by magnificent fires and by day their smoke partly obscured the railroad lines and the bridges over the city's rivers.[50] The artist Aaron Gorson painted nocturnes of the massive steel mills near Pittsburgh that lined the Monongahela River and dominated the surrounding space.[51] The stars are blotted out, and the mills literally define the visual scene as they provide the only source of light. The scale of construction is so large that human beings become invisible. These images celebrate both the built landscape and the dynamism of massive moving machinery. The factory district, viewed from a high place or from a moving train, evoked a sense of powerful forces at work.

As with other forms of the technological sublime, this scene reaffirmed the power of reason, but not in Kant's sense of Reason. Rather than provoke an inward meditation that arrived at a transcendental deduction applicable to all humanity, these landscapes forced onlookers to respect the power of the corporation and the intelligence of its engineers. Many writers and artists saw the landscape of industrialization as something powerful, unique, and unprecedented. Its unnaturalness was part of its attraction.[52] Enthusiasm for the absolute greatness of the constructed landscape had all but blotted out the natural world and provided the illusion of omnipotence.

The contemporary Canadian photographer Edward Burtynsky has developed this tradition of the industrial sublime in enormous photographs of factories, immense oil fields, and open-pit mines whose size and complexity is overwhelming. Burtynsky is aware of the history of landscape imagery, and once declared, "I began by photographing the pristine landscape, but I felt I was born a hundred years too late to be searching for the sublime in nature."[53] He looks "for subjects that are rich in detail and scale yet open in their meaning. Recycling yards, mine tailings, quarries and refineries are all places that are outside of our normal experience, yet we partake of their output on a daily basis. . . . We are consciously or unconsciously aware that the world is suffering for our success. Our dependence on nature to provide the materials for our consumption and our concern for the health of our planet sets us into an uneasy contradiction. For me, these images function as reflecting pools of our times."[54]

In Burtynsky's photographs, as in the industrial sublime of a century ago, the human body is puny compared to the scale of the buildings and excavations. His 2011 exhibition "The Industrial Sublime" presented disturbingly beautiful landscapes that include intense chemical greens, blues, and reds of a mine's tailing ponds, and the carved symmetries of pits seen from hundreds of feet above. Alternately, coal dust may coat an entire landscape, dulling it to grey tones. Burtynsky's industrial sublime presents an unanticipated and unsettling beauty in profoundly degraded and scarred landscapes.[55] Brady, in common with many philosophers, does not accept the technological sublime as a category, and argues that Burtynsky's photographs of mining and other environmental disasters are not sublime but rather fall into the category of "terrible beauty." While stunning as images, in "actual experiences of these places, rather than photographic renderings of them, aesthetic appreciation may become blocked."[56] Yet some tourists and environmentalists do seek out the more shocking and horrifying actuality, as Edmund Burke would have expected.

Where tourists find landscapes sublime, the technologies used to visit them often become so as well. Claudia Bell and John Lyall note, "At the site of the sublime landscape we have technologically sublime people-moving systems, which themselves become signifiers of steepness, danger, and descent." This was early exemplified by steamboats that carried passengers such as Dickens along the base of Niagara Falls. Over time, the "increased velocity and concomitant increased acceleration are more immersed in landscape." They become "full, three-dimensional activations of a sublime event space."[57] A good example is the high-speed glass elevator that rushes tourists to a skyscraper observation deck.

The practice of photography also registered the demand for acceleration. The natural and the technological sublime became common photographic subjects from the 1850s onward. Until the 1880s, taking each image took half an hour or more, because the photographic plate had to be prepared immediately beforehand. The invention of the dry plate sped up the process, but each plate still needed to be inserted, exposed, and then removed. At first the most common images were associated with the wonders located near the new railroad lines, including Yellowstone National Park, Glacier National Park, and the Grand Canyon. After 1910 postcards also depicted automotive destinations. Professional photographers dominated tourist imagery until inexpensive Kodak cameras made photography cheaper, faster, easier, and more common.[58] During this change, the photographer, hidden for decades behind a bulky camera, began to become visible. Time-delay mechanisms made it possible to frame an image from a tripod, press a button, and then hurry into position before the camera snapped the photo. This practice was used more often for group photographs than for self-portraits in sublime landscapes. That changed dramatically with the spread of the smartphone and the "selfie." With it, visitors to the Grand Canyon or the Eiffel Tower could quickly make an image of themselves against a dramatic backdrop and immediately post it on social media.

Rather than dwell on the sublime experience, such visitors dramatize themselves. Many stand on a cliff's edge or by a railroad track with their backs to the abyss or an oncoming train.[59] Holding their smartphone up or mounted on a selfie stick and maneuvering for the best possible shot, some become oblivious to imminent danger. India has many selfie deaths, as do Russia, the United States, and Pakistan. In 2018, the average age of those dying while taking a selfie was twenty-three.[60] Such deaths are under-reported because many are counted as falls, drownings, or other accidents.[61] Amanda du Preez argues that the "contemporary obsession to take an 'epic selfie,' an 'extreme selfie,' or the 'ultimate selfie' may be interpreted as an extension of the pursuit of the sublime." In 2015, people died posing for selfies "taken from the top of a skyscraper while dangling in mid-air, or while perched on the brink of an overhanging cliff just before the selfie-taker's foot slipped." They "no longer look where they are going, but are transfixed by their images on the screen."[62] Many sites ban selfies, but such deaths remain common, often documented by the image of a proud tourist. Such was the fate of a "young Russian girl, Xenia Ignatyeva, who in April 2014 climbed a bridge to impress her friends, but then slipped, fell and was electrocuted by electric fences." It is shocking to see how her "beautiful, young face stares into the camera, exhilarated and energized," unaware that in the next instant she will die.[63] Strictly speaking, such images do not record a sublime experience, for that demands awareness of danger and a sense of fear. A selfie death occurs when the subject is distracted while constructing a dramatic self-image, treating the sublime landscape as a backdrop. A psychologist concluded that "the cell phone pulls you away from the physical environment. You really do tune out the world."[64] For many, the sublime landscape has become a prop.

The technological sublime celebrates progress. In contrast to the natural sublime, which fixes attention on the timeless or eternal, the technological sublime points toward the future. Conquering space, time, and gravity, attention shifts from nature to humanity, from constancy to change, from metaphysics to technique, and in some cases from the landscape to the tourist in the foreground. As the objects of the technological sublime become familiar, however, they lose their luster, and must be replaced, for the technological sublime is future oriented. The Erie Canal was sublime in 1830 but not in 1870; the railroad was sublime in 1840 but far less so in 1890; the telegraph amazed in 1850, but not in 1900; electric streetlights inspired awe in 1880 but not in 1930; the great dams and skyscrapers of the 1930s retain only some of their appeal today. By the 1990s once spectacular technologies were decaying into ersatz spectacles in Las Vegas, including an imitation volcano, casino hotel skyscrapers, a Grand Slam Canyon rollercoaster, and massive lighting arrays.

Yet, while some early forms of the technological sublime have faded into commercialism, ballooning remains popular, and millions of people still pay to visit the observation decks of skyscrapers. New and ever more spectacular structures continue to emerge.[65] The conquest of time, space, gravity, and water draws tourists to visit the world's tallest buildings in the Middle East and Asia, to view the enormous Three Gorges Dam in China, to ride the world's fastest trains in France, Japan, and China, and to observe rockets being launched into outer space.

The achievements celebrated in the technological sublime can fail, however. Inventing the hot-air balloon, railroad, steamship, or airplane made possible spectacular accidents. Constructing magnificent towers, dams, and cities made possible devastating fires, floods, and crashes. As the next chapter makes clear, disasters reconfigure the technological sublime, and its monuments become ruins.

3

DISASTROUS

John Chapin, a reporter for *Harper's Weekly*, was sleeping in a Chicago hotel on October 8, 1871, when awakened by noises in the hallway. He went to the window, "threw open the blinds, and gazed upon a sheet of flame towering one hundred feet above the top of the hotel." He hastily packed and fled with the other guests into the street "to gaze into the face of the awful but sublime monster that was pursuing me." Chapin fled toward the Chicago River, and only then realized that the entire city was on fire and that "no human power could stay its progress." After he was safely across a bridge, he watched the throngs of people escaping with whatever they could carry. "No language which I can command will serve to convey any idea of the grandeur, the awful sublimity, of the scene. For nearly two miles to the right of me the flames and smoke were rising from the ruins and ashes of dwellings, warehouses, lumber yards, the immense gas works." Chapin compared the fire to Niagara Falls. "Everyone knows how inadequate is human language to express the grandeur of Niagara–we can only feel it. And yet Niagara sinks into insignificance before that towering wall of whirling, seething, roaring flame, which swept on, on—devouring the most stately and massive stone buildings as though they had been the cardboard playthings of a child."[1]

The writing on disasters primarily deals with their causes, destructive forces, and human suffering, as well as discussions of who or what was responsible and how to avoid or mitigate future catastrophes.[2] But there is also a sublime aspect to many disasters, as Chapin understood. Both Kant and Burke recognized that earthquakes, fires, tornados, hurricanes, and floods fascinate anyone who sees them. They can be sublime. The philosopher J. Glenn Gray noted, "Anyone who has watched people crowding around the scene of an accident on the highway realizes that the lust of the eye is real. Anyone who has watched the faces of people at a fire knows it is real."[3] Many historical examples confirm Gray's observation. In 1803 the residents of Nottingham in England watched a cotton mill burn to the ground. A journalist lamented the destruction, but he also declared that "no description can do justice to the terrific grandeur of this spectacle." A large crowd gathered as the fire intensified. "The whole of this elegant structure was a prey to this devouring element, ninety windows in front pouring forth columns of flame and combustible matter, so as to endanger haystacks in the meadows at a quarter of a mile distant; and when the roof and cupola fell in, the effect can only be compared to a volcanic eruption. The contrast of this immense and widely diffused light to the darkness of the night, the illumination of the town, castle, and surrounding villages, and the visible distinctiveness of the most distant objects in the landscape, produced such a sublime and vivid effect as it will be as vain for the pencil to delineate as for the pen to describe."[4] In 1840 the fire in a French coal mine presented an even more "dreadful and sublime scene." In a deep circular ravine, fourteen mineshafts gave "access to the innumerable galleries of the mines below." All were aflame and "poured forth with frightful violence from the cauldrons within—flames of a thousand hues rushing forth like whirlwinds–driving, and crossing, and mingling and rising. . . . At times a hollow cracking sound echoes through the abyss" as a tunnel collapsed and sent up "a thick column of black dust" which exploded in "dazzling columns of fire." The whole neighborhood watched the blaze, "blood red roaring and terrible, threatening in its fury to lift up the mountains altogether, and bury the spectators beneath its ruins." Even at midnight, "two thousand spectators are there, some grouped on the opposite crest of the ravine, some sheltered in the cavities of the rocks. Yet no sound meets the ear, save that of the roaring flames." Mute, they had been struck dumb by the terrifying yet "magnificent spectacle."[5]

Burke stated that "there is no spectacle we so eagerly pursue, as that of some uncommon and grievous calamity; so that whether the misfortune is before our eyes, or whether they are turned to it in history, it always touches with delight."[6] Burke discerned two forms: dramatic misfortune "before our eyes" and the ruins of historic disasters. Viewing either form of the disastrous sublime, Burke noted, "is not an unmixed delight, but blended with no small uneasiness." This is especially the case when the catastrophe is recent. "The delight we have in such things, hinders us from shunning scenes of misery; and the pain we feel, prompts us to relieve ourselves in relieving those who suffer."[7] Disasters call forth immediate pledges of assistance, and this response seems as universal as the sublime itself. In the United States, for example, in 1889, large sums were raised for victims of the Johnstown Flood. Money poured in from American states, Great Britain, Germany, and across Europe all the way to Turkey.[8] A similar outpouring of assistance came to San Francisco after its earthquake and fire of 1906. A century later, when a massive earthquake and tsunami in the Indian Ocean killed an estimated 250,000 people, the public response was unprecedented, "not only in the amount of money raised (£10 billion globally), but also in the speed with which the money was pledged."[9]

Burke noted that disasters often are experienced at one remove, through newspapers, paintings, or other representations. Journalists have long recognized that catastrophes sell newspapers, and that if a story bleeds, then it leads a newspaper's front page. Likewise, after circa 1820 panoramas featured disasters. In Paris, *The Burning of Moscow* was depicted in the 1840s. Ancient disasters were popular in Berlin, including a *Panorama of Pompeii* in 79 AD, *The Burning of Rome* in 64 AD, and the Biblical flood. The latter had Noah's Ark in the background, "while in the foreground drowning men cling desperately to uprooted trees, surrounded by a swirling vortex filled with the whole animal kingdom."[10] This panorama of the deluge received great acclaim in 1889. Disaster also lurked in a beautiful German work that depicted the smoking Mount Etna volcano, seen from a village and affording a view of the coastline and the sea. It was displayed at Breslau in 1821, Berlin in 1822, and Vienna in 1831.[11]

The sublime response to overwhelming forces can only occur if one is in relative safety and does not fear for one's life. Like Chapin fleeing his hotel, most of those fleeing the 1871 Chicago fire did not have the margin of safety required to see the conflagration as sublime. But once he was safely across the river, Chapin recognized the fire was an extraordinary spectacle. He was not the only one. William Gallagher, a student at the Chicago Theological Seminary, was awakened by a classmate and told the city was a blaze. He stated that he

> climbed to the roof and saw a sight such as I never expect to see again, and which few men have had the privilege of witnessing. You may read the most vivid accounts of fire that have been written by the most talented men, you may read Schiller's "Song of the Bell" in German, which expresses with great force the power of fire, when it becomes master. You may talk about Moscow and London and New York and Portland fires, but you can never comprehend that single sight, and the constant repetitions we had of it. There was a strip of fire between two and three miles long, and a mile wide, hurried along by a wind . . . sweeping through the business part of this city. We were situated where we could take in the whole. . . . It was a grand sight, and yet an awful one.

When Gallagher went toward the conflagration to help those fleeing with whatever they could carry, he saw few signs of panic. In one prosperous neighborhood, "They were the most philosophical set of people I ever saw. Among those rich people I didn't see one woman rushing about screaming and [w]ringing her hands. There was no crying or bewailing. The very magnitude of the calamity seemed to overcome all those feelings, and everybody set to work to save what they could."[12]

Similarly, the San Francisco earthquake evoked less terror than amazement. The Harvard philosopher William James, who was visiting, found it an exciting event. Even when the room was shaking, he reported, "The emotion consisted wholly of glee and admiration."[13] He later declared he would not have missed it for anything. Many were excited, even ebullient.[14] Twenty-six years after the quake Kathleen Norris recalled, "How I wish that to every life there might come, if once only, such days of change and freedom, so deep and intoxicating a drought of realities, after all the artificialities of civilization."[15] In her view, the earthquake was a welcome natural event that shook people out of their over-civilized lives.

The pipes supplying San Francisco's water had been ruptured by the tremors, and the firemen could not stop the spreading fires. They resorted to blowing up buildings, but even that did not halt the flames. Helpless to save their city, people on higher elevations brought out chairs and sat on

3.1 Residents watch the Great San Francisco Earthquake and Fire of 1906. Arnold Genthe, photographer. Courtesy of Genthe Photograph Collection, Library of Congress, Prints and Photographs Division, https://www.loc.gov/pictures/item/2018704113/.

the street, where they were photographed gazing down at the spreading conflagration (figure 3.1).[16] However terrible the destruction, they realized it was a once-in-a-lifetime moment of awful sublimity. One recalled years later, "The queer feeling of pride at being a witness to the greatest destruction by fire of man's handiwork that had ever been seen by man (Chicago? hm!)." She also recalled "the rarity of evidences of acute grief or despair [sic], but the prevalence of an all-pervading weariness."[17] Those who lived through disasters like the Great Chicago Fire of 1871 and the Great San Francisco Earthquake and Fire of 1906 could never forget their experience of horror and sublimity, followed by exhaustion. But they knew they had seen a terrible magnificence.

Kevin Rozario has argued that the "enthusiasm for disasters was . . . a crucial ingredient of the modernizing process." A "culture of calamity" became an imaginative counterpart to a "social order governed by processes of creative destruction."[18] Viewing ruins of disaster was part of this process. When viewing the broken walls and isolated columns of Chicago in 1871 and San Francisco in 1906 in the aftermath of disaster, or the flood-smashed remains of Johnstown, reporters almost invariably asserted that the will of the people was unbroken and that a new, improved city would rise from the wreckage. *Harper's Weekly* printed a drawing of the first structure erected on the ruins of Chicago: a rough wooden shanty. The sign above the door declared: "All gone but wife, children and energy."[19] This was not an isolated case of bravado. The city's boosters declared that the fire created unprecedented opportunities for investment, and the *Chicago Tribune* editorialized under the headline "Chicago Shall Rise Again."[20] Some even saw the fire as a purification, preparing the city for a central role on the world's stage.[21]

Photographs of ruins circulated widely, including those of Richmond, Virginia, at the end of the Civil War; Chicago in 1871; and central Johnstown in 1889 where the flood had utterly demolished more than 1,600 buildings, and it was all but impossible even to discern the pattern of the streets (figure 3.2).[22] Until the 1880s, professionals took most of the photos, but the democratization of image making came quickly after that, with the introduction of dry plates, roll film, and Kodak cameras. After the San Francisco earthquake and fire, the city's "streets teemed with photographers of all kinds: engineers, geologists, and seismologists preserving photographic data" as well as insurance adjustors, film crews, skilled amateurs with large-format cameras, and hundreds with Kodak cameras. Many came from outside of town. Nick Yablon suggests this enthusiasm for image making shielded people from the trauma of the disaster: "Filtered voyeuristically through the camera's mediating lens, the violent destruction could be distanced or contained."[23] What many saw in postcards, newspapers, and magazines was the disturbing revelation that modern networked cities were vulnerable. Conveniences such as gas lines and electrical systems could endanger an entire metropolis. Poor engineering or shoddy construction meant that a building might collapse in an earthquake or readily burn, spreading destruction to surrounding structures.[24] Ruins of technology might be sublime as spectacle, but they also provided evidence of error, poor workmanship, and political corruption. A disaster that revealed the power of nature might be sublime; shoddy technologies were not. But failure had its positive implications, as it suggested how to do better. As for the calamities of

3.2 The Johnstown Flood of 1889 killed more than 2,209 people and destroyed four square miles of downtown, including almost every wooden building. Courtesy of Library of Congress, Prints and Photographs Division, https://www.loc.gov/item/2005683590/.

nature, as Ted Steinberg has observed, in the late nineteenth century, "the business class in cities throughout America tried to normalize events such as earthquakes by draining them, as best they could, of any larger meaning." Disasters "simply happened and wallowing in the spectacle of life turned upside down or prostrating before God" they declared, "only prolonged the agony." What was needed instead was "a calm, disciplined response aimed at putting things back in order."[25]

One did not need to experience a catastrophe directly to appreciate the disastrous sublime, which was ancient as well as modern. A prime example is Pompeii, a Roman city buried by a volcanic eruption (figure 3.3). In 79 AD a dense cloud of hot pumice, rocks, and fine ashes fell over the city for eighteen hours, collapsing roofs and burying the population under 2.9 meters of debris. The following day, surges of hot gas and pumice swept down the mountain, including two waves of at least 100

3.3 Pompeii, The Forum, with the smoking Mount Vesuvius in the distance. Photo-
chrome print. Detroit Publishing Company, 1905. Courtesy of Library of Congress, Prints
and Photographs Division, https://www.loc.gov/pictures/resource/ppmsca.52663/.

degrees Celsius that covered Pompeii with another meter of rubble.[26] No
one in the town survived. It remained buried until excavations began
between 1748 and 1754. They aroused interest across Europe. One anonymous writer then observed that "empires, however firmly founded, and
that cities, however embellished, are like man, subject to mortality, and
liable to dissolution. This thought naturally humbles the mind in the
dust, and we learn to know our own insignificance, the vanity of our
pretensions, and the futility of all earthly glories." These thoughts on the
insignificance of the individual in the larger scheme of history constitute
the moral lesson the disastrous sublime was expected to teach. The frame-
work for contemplating Pompeii when its ruins were opened to the public included all the elements necessary to evoke a sublime response: the
irresistible dynamic power of nature, a long historical perspective, and

the weakness of humanity. When Goethe visited in 1787, he observed, "There have been many disasters in the world, but few that have given so much pleasure to posterity."[27]

During the nineteenth century, the tide of tourism grew, stimulated by improved transport, the publication of guides, and the rise of the Thomas Cook agency and other travel bureaus. When Charles Dickens visited Pompeii, he thought, "Nothing is more impressive and terrible than the many evidences of the searching nature of the ashes, as bespeaking their irresistible power, and the impossibility of escaping them. In the wine-cellars, they forced their way into the earthen vessels, displacing the wine and choking them, to the brim, with dust. In the tombs, they forced the ashes of the dead from the funeral urns, and rained new ruin even into them. The mouths, and eyes, and skulls of all the skeletons, were stuffed with this terrible hail."[28] Dickens ascended Mount Vesuvius that evening, to view the volcano and the city below in the moonlight. He was on foot, but soon a funicular was built on the volcano's side, making it easier for visitors to gaze into its smoking depths and to view the ruins of Pompeii in the distance.

Those unable to visit Pompeii could see its destruction recreated in dramatic performances. In Britain, the impresario James Pain staged *The Last Days of Pompeii* and then brought it to Manhattan Beach in 1879, well before Coney Island had developed its multiple amusement parks. The sensational effects of Pain's show filled grandstands that seated 10,000 people, and it ran every summer until 1914, even though the Dreamland amusement park constructed a second show about Pompeii in 1904.[29] Pompeii was regularly destroyed in thirty American cities.

As Burke knew, disasters excite the passions and arouse immediate sympathy. Simulations of disasters were common attractions at world's fairs and in amusement parks, including some notorious technological failures. One of the worst was the dam that collapsed and released a wall of water that smashed through the city of Johnstown, Pennsylvania, carrying everything in its path to a stone bridge where desperate survivors were trapped amid broken houses and a great accumulation of debris until the whole mass caught fire, adding a deadly conflagration to the devastation of the city upstream. The Johnstown Flood dominated news headlines for a week and was reenacted at amusement parks for a generation. Coney Island, for example, restaged the Johnstown Flood many times a day for several years.[30] After the 1900 Galveston Hurricane killed between 6,000 and 10,000 people and destroyed most of that city, Coney Island also recreated that disaster starting in 1902. This show proved so popular that it was restaged on "The Pike" at the St. Louis World's Fair of 1904 in a structure built for the purpose that cost $60,000. For twenty-five cents a visitor saw a meticulous dynamic scale model of Galveston before the storm struck, where ships moved, trains crossed the bridge from the mainland, and traffic drove along the streets. A narrator explained that Galveston was a wealthy community and the largest port in Texas. The performance included special effects managed by a staff of fifty, beginning with the gentle plash of ocean waves, the cries of seagulls, and light breezes. The scene gradually darkened, and the city's lights came on. Then the storm gathered and began to batter the city. As towering waves crashed and a powerful, shrieking wind blasted the spectators, they heard actors crying for help. The waves, winds, and powerful tides smashed the city to rubble, illuminated by flashes of lighting. At Coney Island, the performance ended at dawn amid the ruins. But two years later in St. Louis a third act was added, in which the city was rebuilt, including a massive sea wall to protect Galveston from future storms. The performance was repeated all day long, and in St. Louis it made a profit of almost $200,000.[31] The Galveston Hurricane was also staged in Boston, Columbus, Indianapolis, Detroit, Chicago, Milwaukee, Minneapolis, and Kansas City.[32]

Crowds paid to see recreations of disastrous floods and fires at amusement parks, and to see staged head-on railroad collisions between steam locomotives. At one such event in 1896 the crash could be "heard for miles and the air was filled with debris. Both engines sprang into the air and held each other by the nose, while the cars piled up behind them. The wreck was perfect. The locomotives remained forming a perfect Y and all the cars were telescoped. The air was filled with steam and hot water and 25,000 people were panic stricken. When the air cleared and the wreck could be seen, a cheer went up from 25,000 throats."[33]

Lauren Rabinovitz's *Electric Dreamland* presents the culture of disasters as a part of the coevolution of American amusement parks and cinema. "By 1910 every municipality with a population of more than twenty thousand had both amusement parks and motion picture theaters."[34] The parks and theaters recreated disasters that taught Americans to anticipate and even enjoy dangerous events, such as wrecks, floods, or fires. The disastrous sublime offered by amusement parks was a psychological experience that included not just sights but also sounds, winds, and smells that created a powerful simulation of disaster, viewed in comfortable seating. By comparison, early cinema offered only silent black-and-white images with musical accompaniment.

Celebrated disasters frequently combine the natural and the technological. In many cases, natural forces were more devastating because of technological errors. Human beings created a dangerous situation; then natural forces struck. The cities destroyed had been built on low ground beside rivers (New Orleans), on flat islands (Galveston), or downstream from dams (Johnstown) exposing them to flood waters. San Francisco was built on a major faultline where earthquakes frequently occurred. Much of Chicago in 1871 was built of wood and experienced frequent fires. During the twentieth century, societies began to understand spectacular disasters as failures of infrastructure. But during much of the nineteenth century, disasters were generally thought to be "acts of God." To those who read about them in newspapers or saw them reenacted, their awful sublimity was part of an inscrutable destiny. Why some survived and others died seemed a quirk of fate. During the twentieth century, however, human agency in calamities and disasters was gradually accepted. The engineer, the city planner, and the government were held responsible for building the inadequate infrastructure of San Francisco. In the modern age, human beings were expected to control nature. The ruins of New Orleans after Hurricane Katrina in 2005 did not evoke the same response as the ruins of San Francisco a century earlier: "Rather than welcoming feelings of awe and reverence to preserve a place as a national icon, Katrina tours invite outrage, identification, and motivation to help garner redress to transform the material and symbolic power relations toured."[35]

An apocalypse is more than a large disaster. It places a structural limit on sublime experience because it precludes any sense of personal safety. By definition, an apocalypse is such a large-scale event that it exceeds planning or anticipated consequences. Disasters such as hurricanes or volcanic eruptions are immense, but human survival elsewhere can be taken for granted. The forces of destruction are terrifying, but recovery is part of the narrative, as in the later staging of the Galveston Hurricane which concluded with the construction of the seawall around the rebuilt city. But destruction in apocalyptic narratives is so comprehensive that only remnants of society survive, and it is impossible to rebuild much of what has been lost. An apocalypse is approached in Edward Burtynsky's photographs of open-pit mines or oil derricks, which have so destroyed a landscape that it is impossible to return it to anything like its earlier appearance. Jennifer Peebles calls this the "toxic sublime," which emerged after such modernists as Charles Sheeler and Margaret Bourke-White celebrated the machine.[36] Peebles defines the toxic sublime as "the tensions that arise from recognizing the toxicity of a place, object or situation, while simultaneously appreciating its mystery, magnificence and ability to inspire awe." Images or narratives in this tradition encourage viewers to reflect on both sublime devastation and their "position within a polluted world."[37]

Baudrillard's description of "modern demolition" is a suggestive contrast. He enjoyed observing the intentional destruction of a building, using carefully placed charges of dynamite and other explosives. To him, it seemed "truly wonderful. As a spectacle, it is the opposite of a rocket launch. The twenty-story block remains perfectly vertical as it slides toward the center of the earth. It falls straight, with no loss of its bearing, like a tailor's dummy falling through a trap door."[38] No one is inside the building, of course, and its disappearance is a spectacle that typically makes way for erection of a taller building. A demolition is not about loss but about an almost magical disappearance. A demolished building can be reconstructed or replaced, however not every site is recoverable, notably sites with intensive chemical or nuclear pollution, underground coalmine fires smoldering out of control, or lakes that have dried up due to global warming. Only some of these places can be recovered as habitable or useful. If open-pit mines are flooded, for example, they may become recreational lakes. But in other cases, the damage is on such a scale and the accompanying pollution so severe that the site is a fatal environment that cannot be safely inhabited. In an apocalypse, enormous areas may become antilandscapes, and the few humans who survive could hardly see them in terms of the sublime. They must flee, searching for safety.

Yet, temporary toxicity may be sublime. In Don DeLillo's *White Noise*, the main characters must evacuate their home because of an airborne toxic event caused by leaking chemicals. There are three stages to this experience. The first is denial that such a thing can be happening at all. Next comes exposure to the event itself. "The enormous black mass moved like some death ship in a Norse legend, escorted across the night by armored creatures with spiral wings. We weren't sure how to react. It was a terrible thing to see, so close, so low, packed with chlorides, benzenes, phenols, hydrocarbons, or whatever the precise toxic content." Yet confronting this cloud of death "was also spectacular, part of the grandeur of a sweeping event," and their "fear was accompanied by a sense of awe that bordered on the religious."[39] Those who viewed the Great Chicago Fire or who lived through the Great San Francisco Earthquake and Fire of 1906 felt something similar. Like DeLillo's protagonists, they recognized that it is "possible to be awed by the thing that threatens your life, to see it as a cosmic force, so much larger than yourself, more powerful, created by elemental and willful rhythms."[40] The experience of a simulated apocalypse or of an antilandscape is a staple of the film industry. Barry Langford has observed that screened apocalypses "are enjoyable rather than horrifying *because* they are fantasy: in fact, because the technological capacity to deliver a photo-realist simulation of the end of the world reassures us of technology's ability to contain by imagining and acting out any threat, however final and non-negotiable."[41]

Miles Orvell argues that "a new category of visual representation: the destructive sublime" has emerged in recent photography, notably in images of Detroit's decay, New Orleans after Hurricane Katrina, and the destruction of the World Trade Center on 9/11.[42] Like Burke two centuries earlier, Orvell senses both the unease caused by destruction and its attraction. The "framing of disorder" raises "moral and aesthetic questions: is it a betrayal of the chaos of reality, an attempt to make it manageable? Is it a distancing, an aestheticizing of the pain, transforming it into something for our visual entertainment? Or is it a memorial act, an aesthetic rendering that takes us closer to the experience of devastation than we could otherwise go?" Orvell concludes that "a mixed response is inevitable, as one views these images with ambivalence and discomfort and also with an aesthetic thrill."[43]

This analysis suggests why some people seek out scenes of destruction, including the intensely polluted landscapes of the disastrous sublime. There is widespread interest in apocalyptic landscapes. Burtynsky's scenes of waste and pollution draw crowds to exhibitions. There are tours to the Nevada's atomic testing site,[44] and tourists pay guides to lead them into Chernobyl's radioactive zone. One visitor "was on a kind of perverse pilgrimage. I wanted to see what the end of the world looked like. I wanted to haunt its ruins." He saw "modernity in wretched decay" and in "the crumbling ruins of the present" he glimpsed "a world to come."[45] Some seek only the thrill of viewing a fatal landscape, but others want to intensify their environmental awareness, and they expect their guide to provide an educational experience. Goaster and Brunsden suggest that the photography of Chernobyl should be considered a form of "the postmodern sublime."[46] Some sites of dark tourism research, such as the pagan ruins of Cornwall, may not involve the sublime.[47] But other sites qualify, notably the radioactive ghost town of Hanford, Washington, or Picher, Oklahoma, which was so polluted by lead mining that its inhabitants have been moved away.[48] The fascination with comprehensive devastation culminates in visits to environments that cannot sustain human life—and these are often portrayed in books and films depicting a near future where most of humanity has been wiped out.[49] Cormac McCarthy's novel *The Road* is a prominent example.[50]

The imagination of future ruins is not new. As Orvell demonstrates in *Empire of Ruins*, it can be traced back at least to the eighteenth century, including dramatic paintings and photographs as well as fiction, in each case depicting fragments of the present that survive into a dark future. The attraction of this trope seems to be intensifying in the Anthropocene.[51] Joanna Nurmis found that between 2005 and 2015 there was "a strong artistic trend toward the imagery of apocalyptic sublime," especially in works that deal with climate change.[52] For decades, scientists accumulated data on the threat of global warming but were unable to convince the public of the urgency of the problem. The humanities, particularly through visual representations, took up this communication problem. Nurmis analyzes "the eerily attractive spectacle of the sublime, whether a vision of natural beauty about to be lost or a depiction of massive human threats to that beauty." Such spectacles do not necessarily lead to environmental action, however. As Finis Dunaway emphasizes,[53] many "green images" that circulate in the mass media lead not to engagement with environmental problems but to a somewhat illusory green consumerism, such as eating organic beef, which, while it is free of antibiotics and pesticides, requires just as many resources as ordinary beef for its production.

Nurmis argues that art can avoid the pitfall of green consumerism in three ways. First, its representations can heighten awareness and help viewers to engage with environmental problems. Burtynsky's photographs are one example. Second, artists can create installations like Olafur Eliason's work *Your Waste of Time*. He placed 800-year-old blocks of ice from a Greenland glacier in a refrigerated room at New York's Museum of Modern Art. Third, artists can intervene in public space, "to force an encounter that would not otherwise happen." Such works reach a wider public than museum art. A good example is Eliason's construction of a clock made from Greenland ice, placed in front of Copenhagen's city hall. As he put it, "*Ice Watch* makes the climate challenges we are facing tangible."

Some works of "green art" are elegiac or melancholy, and others are overtly apocalyptic, displaying the demise of the landscape due to climate change. Such art provides what Timothy Morton terms "the delight one experiences when recognizing the potential for pain without actually feeling it."[54] Art that portrays how modern technologies such as mining and oil extraction threaten the natural world often does so in an elegiac mode. This sort of "apocalyptic sublime," Nurmis concludes, is "merely providing images and sensory experiences which inspire a sort of catharsis that does not result in any essential reconfiguration of the mind with regard to our relationship to the environment." We can all too easily become voyeurs, enjoying the spectacle of destruction, as in films about the end of the world or in Marc Quinn's exhibition *The Toxic Sublime*, which included large "distorted three-dimensional canvases of seascapes." As one critic summarized, "Each seascape attempts to meld the traditional romanticism of painters like J. M. W. Turner with layers of tape, paint and imprints of drains and debris to illustrate environmental degradation."[55]

The natural sublime of the early nineteenth century understood disasters as a terrible and inscrutable "act of God." The technological sublime of the modernist period presented engineering works as bulwarks against disaster, like the seawall constructed around Galveston. But what form of the sublime is possible in the Anthropocene? One can focus too much on spectacular disasters like the bursting dam that destroyed Johnstown, and miss another sort of disaster, when a new megadam slowly inundates a fertile river valley, destroying its ecology and its farming communities. Rob Nixon argues that megadams in India have produced "developmental refugees" who are rendered invisible by the rhetoric of progress, globalization, and hydroelectric modernity. Anthropologist Thayer Scrudder has estimated that during the twentieth century large dam projects displaced between thirty and sixty million people. They have been written out of the script of progress and become "uninhabitants." Dams more than fifteen meters high did not exist until the twentieth century, when 36,562 were built. Hoover Dam was the exemplar of this movement. It combined "practical purpose and aesthetic ideals by marrying a miraculous feat of American engineering to a sublime spectacle of grandeur." These "megadams became places where the transcendentalisms of religion, nation, science, and art would converge."[56] The sublimity of dams was assimilated to modernization, whether in communist nations led by the Soviet Union, capitalist nations led by the United States, or the postcolonial societies of the Third World. Enormous dams became secular temples symbolizing the conquest of nature. From this perspective, on the one hand, the technological sublime may function as a form of false consciousness, drawing attention away from devastating effects of large engineering projects. On the other hand, the sublime may also be the intended experience of an artwork that draws attention to global warming or species extinction caused by millions of consumer decisions that collectively have unintended consequences: polluted water supplies, islands of discarded plastic in the world's oceans, or the melting of polar icecaps. The technological sublime can become an ideological mask hiding disaster, but it can also be transformed in art to call attention to the slow violence of the disastrous sublime.

These quite different possibilities are particularly evident with atomic technologies. Nuclear bombs did not elicit pride in human achievements, but rather, dread. After Hiroshima, there was an uneasy oscillation between fear of a nuclear holocaust and the hope that nuclear power could provide limitless energy.[57] As nuclear weapons proliferated, however, a new conception of infrastructures emerged that saw cities not in terms of the mastery of nature but rather in terms of vulnerability and ruin. As Stephanie Wakefield notes, in recent times, "Infrastructures have been increasingly recast as bulky and brittle systems incapable of surviving a world of complexity and volatility. The once 'glorious' notion of order promised by infrastructure is now derided as an artifact of an exhausted and imploding humanist era. And modern infrastructure, once thought the pinnacle of the era, is now seen as epitomizing a fatally flawed idea of hubristic human mastery and the cause of today's cascading damage."[58] Resilience may come from the public's improvisation to overcome system failures. In an apocalypse, however, there is no safe place where one can experience the sublime. Infrastructure would fail permanently, everyone would be a victim, and there would be no one outside the apocalypse left to help. (In contrast, when viewing the simulations of apocalyptic films, the public are surrogate outsiders, witnessing destruction in safety.)

Disastrous sublimes have evolved with changes in technology. Eighteenth-century catastrophes were ascribed to fate and described as "Acts of God"; it was unusual in that era to fix the blame on human beings, and few exhibitions or museums were dedicated to disasters. As humanity's technological powers increased, however, disasters kept pace in both speed and scale. Likewise, their representation in panoramas, photography, amusement parks, films, and other media became common. Technological triumphs became potential ruins, and celebrating the sublime was matched by a growing sense of responsibility for both actual and potential failures. As it became clear that earthquakes, fires, and floods had no intentions, human agency became salient, and with agency came accountability.

The scale of disasters is matched by war's devastation.

4

MARTIAL

During World War I German airships, or zeppelins, made fifty-one air raids against Britain. D. H. Lawrence witnessed one of them. "Then we saw the Zeppelin above us, just ahead, amid a gleaming of clouds: high up, like a bright golden finger, quite small, among a fragile incandescence of clouds. And underneath it were splashes of fire as the shells fired from earth burst. Then there were flashes near the ground—and the shaking noise. It was like Milton—then there was war in heaven. But it was not angels. It was that small golden Zeppelin, like a long oval world, high up." The zeppelin was "golden like a moon, having taken control of the sky; and the bursting shells are the lesser lights. . . . the envelope of the sky burst out, and a new cosmos appeared."[1] As is often the case, this description of the sublime disengaged visual impressions from consequences. During World War II, Gray also realized that aerial warfare was aesthetically pleasing. He found that the "spectacle afforded by combat planes is hard to exaggerate. . . . Combat in the skies is seldom devoid of the form, grace, and harmony that ground fighting lacks. There are spectacular sweep and drama, a colorfulness and a precision about such combat which earlier centuries knew only in a few great sea battles."[2] Witnessed in safety, battle could be viewed as a sublime spectacle.

Shortly after watching the first balloon ascent in 1783, Franklin wrote to Joseph Banks in England about the potential military value of "elevating an Engineer to take a view of an enemy's Army, Works, etc.; conveying Intelligence out of a besieged Town, giving Signals to distant places, or the like."[3] Eleven years later the French created the first military ballooning unit, but Napoleon disbanded it after limited service. In the following decades balloonists often entertained the public, but armies were slow to make much use of the technology. Russian balloonists made observations at the siege of Sebastopol. In the United States, President Lincoln saw a balloon demonstration in June 1861. Afterward, the Union Army used observation balloons along both the Potomac and the Mississippi, although communicating with the ground in a timely fashion proved a challenge. During World War I, both sides used tethered balloons, "spotting targets for the artillery and keeping an eye on movement behind enemy lines."[4] Balloonists who hung above the battlefield may have had moments when the vast scene below appeared sublime, but they did not meet a fundamental criterion that Kant set for experiencing the sublime—that the observer be in relative safety. The enemy shot at balloonists and airplanes attacked them. Balloonists wore parachutes and were ready to jump at a moment's notice.[5]

As this example suggests, the martial sublime involves many of the same machines as the technological sublime, but they are employed for quite different ends. The martial sublime has two modes: the experience of combat and the view of military landscapes. The immersive mode can only be experienced in battle, for example, during a spectacular fire fight at night. Martial landscapes are experienced after the fact, for example, by visiting a panorama or a battlefield preserved for posterity.

The perilous magnificence of warfare was well known in the eighteenth century. Burke observed "that the sublime is an idea belonging to self-preservation." He noted that the sound of artillery "awakes a great and awful sensation in the mind," and that "the shouting of multitudes has a similar effect."[6] Hearing such sounds on a battlefield rivets attention to the immediate present to the exclusion of all else, and it heightens the sense of danger. Accounts of battle often focus not on what a soldier could see but on the sounds of artillery, mortar fire, and nearby explosions. One soldier recalled a landing on a Pacific island invasion as "a nightmare of flashes, violent explosions, and snapping bullets. Most of what I saw blurred. My mind was benumbed by the shock of it."[7] A little later, pinned down in a shallow crater, he recalled, "The shells fell faster until I couldn't make out individual explosions, just continuous, crashing rumbles with an occasional ripping sound of shrapnel tearing low through the air overhead amid the roar."[8]

From Homer's time until the nineteenth century, warfare was often described as a combination of strategy and individual heroism. The Crimean War and the American Civil War demonstrated, however, that improved artillery could lay down a lethal rain of death. These also were the first major conflicts depicted in photographs, which showed the dead strewn across battlefields. Matthew Brady sent teams of cameramen to document Civil War battles, and their images shocked contemporaries. One famous image made by Timothy O'Sullivan at Gettysburg depicted heaps of bodies and it became known as "A Harvest of Death" (figure 4.1). Photographs of the Antietam battlefield showed, as one contemporary put it, "what a repulsive, brutal, sickening, hideous thing it is, this

4.1 Incidents of the war. A harvest of death, Gettysburg, July 1863. Timothy H. O'Sullivan, photographer. Courtesy of Library of Congress, Prints and Photographs Division, https://www.loc.gov/pictures/item/2018667213/.

dashing together of two frantic mobs to which we give the name armies."[9] In contrast to most paintings of warfare, such images helped to demystify the supposed glory of military service. Nevertheless, in 1914 many journalists and military experts "envisioned conflicts that turned on brief, decisive battles and heroic deeds."[10] World War I proved to be a bloody standoff, as machine guns lay down a withering fire over the muddy trenches of a no man's land laced with barbed wire. Newly invented tanks and airplanes further mechanized the slaughter. Few soldier accounts of that war describe its battles as sublime, except for descriptions of aerial combat, a last bastion of individual heroism. A generation later, after the firebombing of World War II, after Hiroshima and Nagasaki, one might have thought that the martial sublime had become obsolete, but such was not the case.

After the "mutual assured destruction" of the Cold War, after the spread of nuclear weapons to Russia, Britain, France, China, India, Pakistan, Israel, and North Korea, the idea of a martial sublime might seem a logical contradiction. Moreover, military firepower continues to increase, with missiles, drones, robots, and cyber weapons in the arsenal that can be deployed on electronic battlefields. Logically, perhaps, the martial sublime should have expired after the mechanization of warfare, but it has persisted. A generation after the American Civil War, its horrors seemingly were forgotten, and to some degree this collective amnesia recurred a generation after World War I. Moreover, as Orvell notes, governments have increasingly sought to control the imagery of battle. In the 1940s, much "was shrouded from the general public. Images of dead soldiers, of the seriously wounded, of psychiatric casualties, of the casual atrocities committed by American soldiers in World War II—such images of the brutality of war were censored by the U.S. government and so did not seriously impair a sense of reality based on Hollywood war movies."[11]

During the nineteenth century and after, many tourists visited battlefields, including the notable examples of Waterloo and Gettysburg. Both can still be studied on the ground, and both are memorialized in permanent panoramas. Battles were popular panorama subjects for a century, including an enormously profitable one in London that depicted Lord Nelson's victory at Trafalgar and another showing Lord Wellington's defeat of Napoleon. Between 1880 and 1900 Americans flocked to see panoramas of *The Battle of Chattanooga* (Philadelphia 1885), *The Battle of Manassas* (Washington 1886–1890; St. Louis World's Fair of 1903), *The Battle of Bunker Hill* (Boston 1888), *General Custer's Last Fight against the Sioux Indians* (Boston 1889; Montreal 1892), *The Battle of Manila* (Philadelphia 1899), and *The Battle of Gettysburg* (Boston 1883). *The Battle of Gettysburg* toured to Newark, New York, Baltimore, and other cities before finding a permanent home adjacent to the Gettysburg Battlefield. Battlefields remain popular tourist destinations, whether the trenches of World War I, the beaches of D-Day, or the ruins of Pearl Harbor.[12]

Another expression of the martial sublime is vast ranks of soldiers marching. In 1859, some American tourists were deeply impressed when by chance they passed through the columns of twenty thousand French troops marching rapidly through a snowstorm in the Alps to fight in Italy. The tourists wrote enthusiastically to the *New Haven Register*: "What a strange circumstance that we should have been permitted to witness the sublime sight."[13] More commonly, civilians saw armies parading in cities. General Sherman concluded his memoir of the Civil War "with the victory parade of the Army of the Potomac." As James Dawes explains, "The parade as an exercise in the sublime is both a performance of the army's seamless unity and a breakdown and analysis of its component parts; it is an imagistic retelling of the story of the war that leads, as naturally as a straight line, to the inevitable end point of its victory."[14] The vast ranks of marching soldiers demonstrated the power of the military, the survival of the Union, and the war's triumphant conclusion.

More spectacular than a parade is an air show orchestrated to entertain and awe the public. In Britain, for example, once each summer between 1920 and 1937, "dozens of military aircraft soared and dived over Hendon" in North London, where they "looped and tumbled, flew in formation and fought mock battles, before huge, enraptured crowds." Royalty regularly attended, and crowds exceeded one hundred thousand. They witnessed set pieces, which included bombing "savages" in remote parts of the empire and the rescue of white refugees by aerial transports. But imperialism was a minor theme, as the Royal Air Force (RAF) focused more on industrial targets such as ports, power stations, and submarine bases, and on European threats, particularly from Germany. The crowds were awed by the novelty of flight itself and thrilled by the loud engines, the diving planes, and the danger of accidents. As Brett Holman emphasizes, "the highly technological spectacle of flight made it an almost ideal form of technological sublime."[15] RAF shows were organized into dramatic stories that suggested the experience of battle. Such air shows are still used in the demonstration of new weapons to arms buyers. Politicians, generals, and potential customers witness low-flying bombers traveling faster than the speed of sound, dropping powerful bombs with precision.

Observing an actual battle is a more powerful and immediate experience than is possible through photographs, panoramas, film, museums, curated battle sites, parades, air shows, or weapon demonstrations. Since about 1990 a new class of tourist has emerged who want to visit active war zones. These tourists pay to visit rebels in the mountains of Pakistan or in the Syrian civil war. They photograph burned-out Russian or American tanks or pose before smashed statues in Afghanistan. They pay to watch the Kurds fighting around Mosul or to gaze down from the Golan Heights as Syrian rebels attack Assad loyalists.[16] Some also want to experience reenactments of war, like those who spend a night on the floor of a Sarajevo bunker, disturbed by recordings of gunfire.[17] Such tourists want more reality than a panorama or film can offer, but they still have only a vicarious thrill.

Warfare affords soldiers more powerful experiences of the martial sublime. They must obey orders that put them in harm's way, but they may view spectacular scenes. A British officer enjoyed the successful bombardment of Algiers in 1816, in which every enemy ship was destroyed. When the battle ended at 9:00 p.m., "all that floated of the Algerines was in a glorious blaze, which presented the most awful and sublime scene that imagination can conceive. The night itself was pitchy dark; but such was the effect of this grand illumination, that noon, with its most brilliant sun, could not have displayed a more effulgent light." He concluded by referring to what was then the new phenomena of the panorama. "What a charming panorama it would make with Barker's skill!"[18] The scene immediately before battle can also be sublime. A famous example is the enormous flotilla that landed on the beaches at Normandy on D-Day, June 6, 1944 (figure 4.2). Similarly, in *The Warriors*, Gray recalled "most vividly my feelings while watching from a landing boat, on the morning of August 25, 1944, the simultaneous bombardment of the French

4.2 Bird's-eye view of landing craft, barrage balloons, and Allied troops landing in Normandy, France, on D-Day, June 6, 1944. U.S. Maritime Commission. Courtesy of Library of Congress, Prints and Photographs Division, https://www.loc.gov/pictures/resource /cph.3c11201/.

Riviera by our planes and by our fleet of warships."[19] Gray was temporarily in a safe position on an offshore landing craft, where "thousands of us watched motionless and silent, conscious that we would be called upon to act only after the barrage and bombing were over." In the relative safety that Kant considered a prerequisite for an experience of the sublime, they "saw the planes, appearing from nowhere, and in perfect alignment over their targets. Suddenly, fire and smoke issued from huge cannon on our ships, and the invasion had begun. Our eyes followed the planes as they dived into the melee of smoke and flame and dust and emerged farther down the coast to circle for another run. The assault of bomb and shell on the line of coast was so furious that I half expected a large part of the mainland to become somehow detached and fall into the sea." Gray knew the bombardment meant "havoc and terror" for the local inhabitants, but nevertheless he recognized that "the scene was beyond all question magnificent." He could "gaze upon the scene spellbound, completely absorbed, indifferent to what the immediate future might bring," and his fellow soldiers were equally engrossed. During World War I Siegfried Sassoon was similarly impressed by the "angry beauty" of the battlefield with the artillery flashing like lightning followed by a thunderous roar and punctuated by high-pitched gunfire.[20]

Gray concluded, "Some scenes of battle, much like storms over the ocean or sunsets on the desert or the night sky seen through a telescope, are able to overawe" an individual, who, "lost in the majesty" finds that the "ego temporarily deserts him, and he is absorbed into what he sees." During "these rapt moments" a soldier finds that "the chief aesthetic appeal of war surely lies in this feeling of the sublime."[21] Gray emphasized that this was not a feeling "of triumph, but, on the contrary, a recognition of power and grandeur to which we are subject. There is not so much a separation of the self from the world as a subordination of the self to it." This form of the martial sublime is closely related to Kant's mathematical sublime, in which the individual is also humbled. It does not lead to depression. On the contrary, Gray notes, "We are able to disregard personal danger at such moments by transcending the self, by forgetting our separateness."[22] As Gray observed his feelings and those of his fellow soldiers, he concluded that "the distinctive feeling of the sublime is its ecstatic character, ecstatic in the original meaning of the term, namely, a state of being outside the self" during which a "pervasive sense of wonder satisfies us because we are assured that we are part of this circling world, not divorced from it. . . . we gain a relationship to something greater than the self."[23] This is not the same thing as what Kant meant by Reason. As Melissa Vera Licht argues, "The unifying moment of Gray's sublime is not elevated by Kantian reason; it remains on the level of material realities of life and death. In this way, Gray removes the sublime from the rigor of the moral law which it expresses in Kantian theory."[24]

Scenes of warfare may have this effect, yet by no means do all soldiers share such experiences. As John Keegan emphasizes in his classic work *The Face of Battle*, warfare is chaotic and even generals may have difficulty getting an overview. The common soldier often has no grasp of the larger picture. From the lower ranks "the view was a very local one" of "almost indecipherable chaos."[25] Yet overall, there is a sense of extraordinary vividness, of living intensely in the perilous present. As Harvie Ferguson argues in a seminal article, "The sublime of combat is forcefully, overwhelmingly and inescapably actual. Combat is sublime immanence. It is, nonetheless, the antithesis of the mundane or ordinary world; it is the contradiction of an actual sublime; a paradoxical non-transcendental Otherness."[26] An American officer who fought the British near Niagara Falls in 1814 had a particularly intense experience of this kind. His unit captured the enemy artillery and then resisted "five desperate charges" as the British attempted to recover their weaponry. The "action raged with unexampled fury" as the Americans repeatedly repulsed them. "A more awful and sublime scene the imagination could not have pictured. The deadly roar of Niagara (one of the wonders of nature), the darkness of the night, still darkened by the columns of smoke, the vivid flashes of musketry, the tremendous bursts of cannon, mingled with the cries of dying and wounded" together "conspired to render this an awful scene of horror."[27] Such overwhelming chaos left no room for transcendental thoughts.

In *The Critique of Judgment* Kant suggested a more exalted aspect of war: "Even war has something sublime about it if it is carried on in an orderly way and with respect for the sanctity of the citizens' rights. At the same time, it makes the way of thinking of a people that carries it on in this way all the more sublime in proportion to the number of dangers in the face of which it courageously stood its ground."[28] In the "orderly" warfare that Kant had in mind, civilians would be safe from plunder, rape, or mistreatment, and armies would treat prisoners with respect. Kant would not have considered the use of poison gas in World War I or atomic bombs in World War II to be weapons that "respect the sanctity of the citizens' rights." He considered war a potential stimulus to selfless good conduct.[29]

Most warfare today does not provide the conditions Kant imagined, namely a defense of one's country along well-defined lines of battle, with mutually agreed rules regarding treatment of prisoners and civilians. Such conditions are rare during wars such as those in Vietnam or Afghanistan, where front lines remained ill defined. Ambush, surprise, and other guerilla tactics were common. Many civilians were forced to take sides, and soldiers often could not distinguish between the enemy and those they were defending. Sebastian Junger spent time with U.S. ground troops in Afghanistan and observed long periods of waiting, with no enemy visible, punctuated by sniper bullets and brief but intense firefights. These conditions did not provide much opportunity for sublime experience.[30]

Burke noted, "When danger or pain press too nearly, they are incapable of giving any delight, and are simply terrible." This is the case with a long bombardment or a charge against intense enemy fire.[31] Burke argued that sounds by themselves could produce powerful sublime effects. "Excessive loudness alone is sufficient to overpower the soul, to suspend its action, and to fill it with terror. The noise of vast cataracts, raging storms, thunder, or artillery, awakes a great and awful sensation in the mind."[32] Prolonged attacks became more common after the industrial revolution made possible the mass production of powerful, rapid-fire weapons. The sonics of warfare changed, as the size and explosive power of the shells increased, and more intense barrages became common. In the American Civil War, during the siege of Mobile, Alabama, Union batteries bombarded the Confederate forces with thousands of eight- and ten-inch shells. The ceaseless explosions were deafening. As Major-General C. C. Andrews recalled, "The fire of so many large guns, and the loud explosion of shells, produced one of those sublime scenes which seldom occur, even in the grandest operations of war. There is scarcely anything in the phenomena of nature to which it could be compared; certainly not the distant murmur of the thunder, nor its near and startling crash." The closest comparison might be a thunderstorm in the mountains, when "the dark clouds seem to linger on the mountain-tops, and from all quarters of the heavens the awful bolts burst forth simultaneously."[33]

As the siege of Mobile suggests, there is "a thoroughgoing metamorphosis, in combat," once "hearing rather than vision becomes the dominant sense."[34] Often one could no longer see the enemy but only heard firing, the scream of shells in flight, and their explosions on impact. Consider this description of the sounds during the World War I Battle of Verdun: "Now and then the ear was utterly dazed by a single, absolutely hellish crash accompanied by a sheet of flame. Then an unceasing and sharp swishing gave the impression again that hundreds of pound weights were flying after each other through the air with incredible velocity. Then came another dud, plunging with a short, heavy thump that shook the solid earth all round. Shrapnel exploded by the dozen, as prettily as crackers, scattering their little bullets in a heavy shower, with the empty cases whizzing after them. When a shell went up near-by, the soil rattled down in a torrent, and with it the jagged splinters as sharp as razors rent the air on all sides."[35]

Long before such mayhem was possible, Burke had noted that "when at any time I have waited very earnestly for some sound, that returned at intervals, (as the successive firing of cannon), though I fully expected the return of the sound, when it came, it always made me start a little; the ear-drum suffered a convulsion, and the whole body consented with it."[36] Tension increased with successive explosions, and did not abate when the firing stopped, for "the organs of hearing being often successively struck in a similar manner, continue to vibrate in that manner for some time longer."[37] Two centuries after Burke's observations, studies made during the Vietnam War found that troops have high levels of stress (measured by cortisol levels in blood tests) when waiting for an attack. In contrast, they experienced "euphoric expectancy" and falling cortisol levels when an attack occurred. The men "were more at ease facing a known threat than languishing in the tropical heat facing an unknown one."[38] Compare this to a classic experience of the natural sublime, in which there is no enemy, but rather a powerful experience enjoyed in relative safety. The sights, sounds, and smells endured by troops under attack arouse their senses to a pitch of painful intensity. This alertness, accompanied by a higher blood pressure and accelerated heartbeat, comes at a cost, as "abilities to reason and perceive our surroundings deteriorate. Cortisol interferes with the part of the brain that handles complex thinking. We suddenly have trouble solving problems, even simple ones—like how to put on a life jacket."[39]

In Kant's classical sublime, one is astonished but does not lose the ability to think clearly. But in battle, the martial sublime emerges amid a rush of unexpected, violent sensory experiences. Any one of the bomb blasts, machine gun bursts, or tracer bullets would be enough to rivet one's attention, but when such deadly fire is incessant, when the ground shakes, when the smell of napalm fills the air, when soldiers are wounded, dying, and cry out in pain, the simultaneous impressions overwhelm the senses. There is no Kantian Reason when under siege. As Melissa Vera Licht succinctly summarized, "The possibility of finding meaning in modern warfare seems to be undermined by the lethality and destructiveness of its methods, but this does not kill its appeal."[40]

In its most extreme form, the martial sublime spills over into trauma, as the mind is overwhelmed and afterward cannot assign a meaning to the experience and reach closure. Licht analyzed the power of warfare "to overwhelm and shock the soldier's sensibility" and found that the "sublime aspect of war" marked them. "The sublime elements of war reshaped their view of world and self. It didn't just change their *opinions*, it changed *them*." It did so because "warfare's opposition and vacillation between material destructiveness and absolute idealism generate a negative sublime, which exceeds subjective faculties and leaves them traumatically out of balance."[41] As Philip Caputo said of his war experiences: "Anyone who fought in Vietnam, if he is honest with himself, will have to admit he enjoyed the compelling attractiveness of combat. It was a peculiar enjoyment because it was mixed with a commensurate pain. Under fire, a man's powers of life heightened in proportion to the proximity of death, so that he felt an elation as extreme as his dread. His senses quickened, he attained an acuity of consciousness at once pleasurable and excruciating. . . . it made whatever else life offered in the way of delights or torments seem pedestrian."[42]

Novelist Tim O'Brien, who also served in Vietnam, explained, "For all its horror, you can't help but gape at the awful majesty of combat. . . . It's astonishing. It fills the eye. It commands you. You hate it, yes, but your eyes do not."[43] The physical danger and the overwhelming impact of modern battle make it difficult for the mind to move beyond sensory impressions. As Licht points out, "The Kantian paradigm of sublimity, in which sensory shock and passivity can lead to active rational reflection, falls apart as shock is followed by shock, or by rationalization rather than reflection."[44] Yet paradoxically, the martial sublime results from human reason and ingenuity. A war is made possible by thousands of meticulously planned efforts, including the mining of metals, their manufacture into weapons, the training of troops to use those weapons, and the deployment of the army. However, despite intensive application of human reason to every aspect of war, troops know that battles may result not in victories but in chaos, destruction, defeat, and death.

Well before the Vietnam War, the awe of the Kantian sublime had been intentionally weaponized in the planning for aerial warfare. The Dutch inventor Anthony Fokker declared in 1922 that "the airplane of the future will be at once the most ghastly and sublime machine ever created by the hand of man." Giulio Douhet had a similar realization. His *The Command of the Air*, published in Italian in 1921, influenced Mussolini whose air force attacked Ethiopia in the 1930s. Douhet argued that future wars would be won through precision bombing of an enemy's factories and supply lines, which would cripple the army and demoralize civilians. He declared that "the more rapid and terrifying the arms are, the faster they will reach vital centers and the more deeply they will affect moral resistance."[45] His book was required reading for officers of the U.S. Army Air Corps and influenced their bombing strategy during World War II. Similar ideas later emerged in the "shock and awe" tactics used during the American war against Iraq. Licht perceptively notes, "The campaign's title turns on the two moments of the Kantian sublime: shock at a massive power, awe at moral and rational law. Aerial bombing exacts a grotesque transformation of this progression, aligning supposed moral righteousness with vast physical destruction—precisely the terms Kant attempted to place in tension. This collapse of meaning into aggressive force and violence is a powerful distortion of aesthetic experience."[46]

Douhet was wrong about how effective bombing would be and about its psychological effects. He expected that intensive aerial bombing would pulverize a city, demoralize its citizens, and undermine their will to fight. This was an *imagined* sublime on the part of those planning an attack. Douhet and those he influenced had fantasies of domination through extreme violence. However, they overestimated the accuracy and the destructiveness of the bombs and underestimated the civilian response. When the citizens of London were subjected to intensive bombing during World War II, the city continued to function.[47] Bombing did not demoralize the British but aroused patriotic anger and increased their will to resist. Similarly, the idea that bombing could awe the Vietnamese or the Afghan peoples into surrender proved an illusion. The predictions about human behavior by military strategists who advocated "shock and awe" bombing were misguided. However, as Tanine Allison has shown, films made after World War II often presented explosive violence against enemy targets as a form of the "destructive sublime" to be retrospectively enjoyed by the victors.[48]

In Europe, the destructive sublime of World War II culminated in the firebombing of Coventry in England and Dresden in Germany.[49] Images of their destruction resembled earlier photographs of cities in ruins. From Hiroshima and Nagasaki, however, came not only photographs of broken stone buildings and wooden houses reduced to ashes, but also new unsettling images, such as a schoolgirl's watch with the hands melted into the dial at the precise time the first atomic bomb exploded (figure 4.3), or the shadow of a man next to a ladder "printed" on the side of a building by

4.3 A schoolgirl's watch, marking the time when the atomic bomb exploded over Hiroshima, and its blast of heat incinerated her and melted its dial. Hiroshima Peace Museum. Photograph from Wikipedia Commons.

the heat of the atomic blast. As Colem Hemez has noted, the shadows of bodies "scorched into the concrete surfaces of the city" demonstrate that "the sheer energy emitted by the bomb circumvents the need for a chemical mediator localized within the photosensitive layer of photographic film. The atomic bomb seems to turn the entire surface of the Earth into a photographic plate."[50] Such horrific images were circulated by pacifists and anti-war activists. They forced one to realize that the blood of anyone exposed to an atomic bomb explosion instantly boiled and vaporized as their bodies disintegrated.

In contrast, the government and much of the press emphasized distant views of the enormous clouds that atomic bombs created. Peter Bacon Hales analyzed the "emphasis on natural imagery" in descriptions and photographs of early atomic blasts. "By choosing such analogies, the writers did more than simply appropriate a language that could illuminate this new phenomenon. They bridged a previous gap between what was human and what was natural—the atom bomb became a man-made marvel of nature, and thereby the question of responsibility for the effects of the explosion remained slippery."[51] Many of the first images of the bomb made it look like a tornado or a storm. At the same time, much early journalism about atomic bombs presented them "as a continuation of modern warfare" rather than as something fundamentally new. The only journalist on the plane that dropped the atomic bomb on Hiroshima was William L. Laurence of the *New York Times*. He presented the bomb in aesthetic terms, without reference to the human devastation it caused, depicting it as something alive.

Awestruck, we watched it shoot upward like a meteor, ever more alive as it climbed skyward through the white clouds. It was no longer smoke, or dust, or even a cloud. It was a living thing, a new species of being. . . . At one stage it assumed the form of a giant square totem pole, with its base about three miles long, tapering off to about a mile at the top. Its bottom was brown, its center was amber, its top white. . . . It was as though the decapitated monster was growing a new head. As the first mushroom floated off into the blue, it changed its shape into a flowerlike form, its giant petal curving downward, creamy white outside, rose-colored inside. It still retained that shape when we last gazed at it from a distance of about 200 miles.[52]

The focus on the sublime cloud rather than the effects of the bomb was common in the writings of military officials, scientists, and journalists at the time, and "the atomic explosion became not a purely human circumstance (for which we must accept responsibility), but rather a part of that benign collaboration among man, nature and divinity that had defined American destiny."[53] There was seldom a word about the Japanese incinerated in the blast.

Likewise, there was little awareness of radioactive poisoning, and in Nevada during the 1950s thousands of troops were ordered to view open-air tests, often crouched in trenches close to the blast.[54] Thousands of tourists and local residents also ascended nearby hills to watch. Those living nearby, called "downwinders," later had a disproportionate number of children with birth defects, which they called "sacrifice babies."[55] After the end of open-air atomic tests in 1963, the atomic sublime was evoked at historic sites, notably at the Nevada Test Site (NTS) where 1,054 of America's open-air and underground atomic tests were held,[56] and at museums, including one in Las Vegas affiliated with the Smithsonian that houses the NTS collections. The Department of Energy scripted the narrative that organizes the "eight thousand square feet of exhibition space . . . explaining the massive technological achievement involved in building and testing nuclear weapons."[57] The bombs are presented as physical objects or as blasts seen from a distance with powerful, aesthetic effects. There is little about the civilians who died from the attacks or the effects of radiation. Tourists can also go to the White Sands Missile Range in New Mexico where the world's first atomic bomb was exploded. Most of the 600,000 acres of this still-active test site are off limits, restricting visitors to an outdoor viewing area and a museum. It displays bombs from throughout American history, explaining when they were invented and how long they were used. This story culminates with atomic weapons and the Trinity Test Site.

When the artist James Russell Gowans visited, he found that taking photographs was restricted to either the bombs on display or to an "enormous mountain range," and he "stood there and experienced both drawing my attention, sensing their similarity and difference," as two manifestations of the sublime. He concluded that "the technological sublime is a measure of ourselves not against nature, but against technological and scientific innovation. We hold humanity's intellectual abilities against humanity's own abilities of empathy, our desires for power, and capacities for greed. In the technological sublime our inner powers are always held in contrast to our weaknesses."[58] At the White Sands Missile Range, he "felt insignificant, small, and trapped within a state apparatus that I had no control [over] . . . and engaged with a moment of self-reflection realizing my own position within these systems."[59] Such reflections hardly resemble those of someone witnessing the natural sublime.

Museums, panoramas, amusement park recreations, and films have prepared audiences to aestheticize destruction. But is aesthetic distance possible in the face of a terrorist attack? The traditional sublime assumes the viewer is in relative safety from the storm, fire, or flood, but terrorists intend to abolish any sense of safety and replace it with overwhelming fear. Some might argue that terrorists themselves experience an emotion akin to the sublime. However, the sublime has never been based on premeditated murder or wanton destruction. The German composer Karlheinz Stockhausen called the 9/11 attack "the greatest work of art imaginable." He thought that it epitomized "minds achieving something in an act that we couldn't even dream of in music, people rehearsing like mad for ten years, preparing fanatically for a concert, and then dying; just imagine what happened there. You have people who are that focused on a performance and then 5,000 people are dispatched to the afterlife, in a single moment."[60] He saw 9/11 from the perspective of the terrorists and admired their dedicated preparation rather than viewing the attack from the perspective of the victims and involuntary spectators. Stockhausen conflated mass murder with an avant-garde happening.

It is wrong-headed to see premeditated murder as a work of art, but Stockhausen inadvertently pointed to how the media presented the attack on the World Trade Center from a distance. What most of the world repeatedly saw on television was enormous and almost abstract: large airplanes smashing into the two tallest buildings in New York City, seen from a position of relative safety. Just as Gray recognized the aesthetic appeal of some scenes of war, when seen from a distance; just as the atomic bomb was characteristically depicted as an enormous mushroom cloud, seen from a distance; so too 9/11 was repeatedly reduced to the moment of impact, when a passenger jet slammed into a tower, at the moment when one of the towers collapsed, or as ruins. The human beings on the plane or in the building remain implicit. Orvell has noted that the events of 9/11 may be the most photographed event in human history, but only a few images are widely circulated. Photographs of the identifiable dead, whether from D-Day, Hiroshima, or 9/11, are seldom reproduced. They are not sublime but rather reveal suffering and death. Photographs of those who jumped from the Twin Towers and fell to their death were prominent in newspapers on September 12, but "they quickly disappeared after their initial exposure."[61] The American press, the government, and much of the public preferred the epic, panoramic view.

The 9/11 attacks prompted the Bush administration to retaliate, first in Afghanistan and then in Iraq. These wars were extensively filmed and photographed, but the Pentagon was able to shape both the images and the meanings given to them to a considerable degree. The Bush administration recognized that journalists saw warfare as skeptical spectators, and in response it developed the strategy of embedding reporters in combat units. Now that they were constrained to follow and move along with specific troops, their stories changed. They presented the powerful immediacy of immersion in battle, with little overarching interpretive narrative. The result was what François Debrix describes as "images or objects that, at first glance, shock or do not make sense" but which could be used to justify "a greater idea, concept, or mode of rationalization."[62] Embedded journalism did not invite reflection but rather justified American military actions. The philosophical reflections that are part of the natural or technological sublimes were rarely possible. For as Ferguson emphasizes, modern combat "is a world unlike any other, a world of confusion and chaos, a kind of anti-world. Its sublime Otherness is made real for us, above all, as the negation of order. On the battlefield Chance rules supreme."[63] The preserved landscapes of the martial sublime recuperate a sense of logic and order after the battlefield experience shatters meaning. Through paintings, photographs, parades, preserved battlefields, and reenactments, society restructures the memory of war into a coherent historical event.

One of Hemingway's characters, Krebs, discovers this relationship after returning home from World War I by "reading a book on the war. It was a history, and he was reading about all the engagements he had been in.

It was the most interesting reading he had ever done. He wished there were more maps. He looked forward with a good feeling to reading all the really good histories when they would come out with good detail maps. Now he was really learning about the war."[64] As this example suggests, spectacular sublimes, which are based on unrepeatable and often chaotic experiences, may be recomposed as panoramas, maps, photographs, preserved battlefields, and museums. Chaotic lived experience becomes a static memory. In this way, even an atomic blast can be stabilized as a logical outcome of history.

The contemporary photographer Simon Norfolk defines battlefields broadly to include regions and times beyond an immediate conflict. He describes his aesthetic as a "military sublime" that is "driven very much by an understanding of Romantic philosophy and a vision of 'the sublime'" where "ideas of beauty and horror" are brought together "to produce something exciting and frightening and terrifying, but also uplifting." His images expand "the meaning of the word 'battlefield'" to include "places that have been ruined by war, or created for the purposes of war." These include refugee camps, blasted buildings, and other places of displacement and devastation beyond battle sites. "Moving the idea of 'battlefield' across space and time, and into places you don't quite think of [as battlefields], is at the centre of my work."[65] Norfolk's military sublime addresses the consequences of warfare revealed in landscapes. Referring to enthusiastic announcements of new weapons, discussions of their deployment and use, and videos made in battle zones, he felt that "one cannot be left with anything but conflicted feelings. The bewildering beauty of what human ingenuity can achieve when given endless resources collides with the appalling disposal of those assets on new and more brilliant ways to kill people." His military sublime juxtaposes ingenious means with horrific ends. It is a contrast to the battlefield tourist's sense of awe when reconstructing the organized mayhem in which soldiers died by the thousands, or what O'Brien called a soldier's amazement at "the awful majesty of combat. . . . It's astonishing. It fills the eye. It commands you."[66] Norfolk insists on moving beyond that moment and the narrowly defined space of combat to include war's larger effects of displacement, destruction, and death. Likewise, the atomic bomb dropped on Hiroshima is best comprehended not only as an enormous cloud but also as radiation sickness, shadows of evaporated bodies printed on walls, and a melted watch, its hands pointing to the moment when the blast boiled the blood in an arm and instantly turned that human being into ashes. The martial sublime celebrates battle at the risk of dehumanizing its victims.

This chapter completes the discussion of directly observed, tangible, sublime experiences, whether natural, technological, disastrous, or martial. The next three chapters examine forms of the sublime that require the aid of mediating technologies.

MEDIATED SUBLIMES

5

INTANGIBLE

Most writing on the natural sublime assumes that the objects calling forth this emotion are tangible, that they are visible to the naked eye, and that they impart strong sensory impressions, such as the roar of the Niagara Falls, the howling of a hurricane, or the stunning silent void of a vast canyon. But there are far less tangible sublime experiences as well, many of them connected to scientific research. Already in the seventeenth century inaccessible objects were being revealed by using the telescope and the microscope. In astronomy and biology and later in other sciences, mediated experiences of sublimity became common. Often, these were only accessible through one sense, typically sight. Many intangible sublime experiences could be enjoyed by anyone—for example, viewing the mountains and craters of the moon through a telescope. But other mediating technologies required training to use them as well as education to understand and interpret the objects revealed. These experiences demanded intellectual preparation as a precondition, in contrast to the tangible, often unexpected, confrontations with the natural sublime. The experience of the intangible sublime is not an encounter with immediate danger or with an overwhelming object in nature, the built environment, disaster, or battle. Because these impressions are scarcely palpable and because the apprehension of immediate danger is absent, compensatory knowledge may be needed to attain the sensibility needed for this sublime experience. Some technologically mediated forms of the sublime may only be intelligible to specialists.

The concept of the intangible sublime therefore raises the question of to what extent it demands specific preexisting knowledge. What some philosophers call "scientific cognitivism" maintains "that scientific knowledge is necessary for aesthetic appreciation, that is, that we need knowledge of sciences such as geology, biology, and ecology in order to engage in appropriate aesthetic responses to nature's actual aesthetic properties."[1] In reply, Robert Clewis has argued that "scientific knowledge 'enhances' the free aesthetic appreciation of nature," but such knowledge is not required to experience the sublime. For example, expertise in geology is not necessary to experience the Grand Canyon as sublime, nor need one be an astronomer before photographs of the lunar surface become sublime. Instead of being a requirement, scientific knowledge is better understood as a form of enhancement, an "adherent sublimity."[2]

Nevertheless, some mediated experiences that are difficult to decipher or that are not widely accessible might be disqualified from being sublime. The hallmarks of the sublime are that it is astounding, irresistible, and immediately recognized by all.[3] The data sent back to Earth by a satellite does not satisfy this definition. But modify this requirement slightly, to say that experiences of the intangible sublime must be accessible to all who are willing to try to understand them, and then many areas of scientific research do qualify. For example, the layperson cannot master the details of Einstein's theory of relativity, but nevertheless can grasp the idea of a black hole in space whose gravity is so strong that not even light can escape its pull. When such examples of the intangible sublime are elegantly presented, they meet the requirements that Longinus laid out for the rhetorical sublime. Not only scientists but also millions of others have viewed and understood striking images of things that cannot be apprehended by the unaided senses.

Like other formations of the sublime, this one also comes in two modes. Since both are mediated by machines, the difference between these modes is not as great as that between being a soldier in battle and visiting a battlefield years later. The fundamental difference between the two modes of the intangible sublime is the degree of immersion (or interactivity) in the experience. One extreme is looking at a static image of the moon's surface; at the other extreme is driving a lunar rover on the moon. Viewing an image is the more common experience. The dynamic and interactive mode, like the experience of battle, is only available to those who fly or drive a vehicle by remote control as they explore an unknown place, for example at the bottom of the ocean or on the surface of Mars.

The intangible sublime has appealed to a popular audience since the eighteenth century, and it has featured in scientific demonstrations, public lectures, and publications. Joseph Priestley (1733–1804) used the terminology of the sublime to discuss astronomy and chemistry, which he thought "exhibit the noblest fields of the sublime that the mind of man was ever introduced to." For Priestley, who discovered oxygen, scientific discovery evoked astonishment, and when teaching he believed that "curiosity and surprise . . . should be excited as soon as possible."[4] Priestley was characteristic of his age. As Alan Gross explains in *The Scientific Sublime: Popular Science Unravels the Mysteries of the Universe*, the eighteenth-century interest in the natural sublime spread to the sciences, notably in Adam Smith's *History of Astronomy*.[5] Smith described the psychological effect of a spectacular natural event, such as an eclipse of the sun, during which observers "in vain look around all their classes of ideas in order to find one under which it may be arranged. They fluctuate to no purpose from thought to thought" and this mental activity excites "the sentiment properly called *wonder*, and which occasions that staring and sometimes that rolling of the eyes, that suspension of the breath, and that swelling of the heart" that arise "when wondering at a new object."[6]

A new object could be miniscule. Using the microscope Antonie van Leeuwenhoek (1632–1723) opened a miniature world to observation. Among his many discoveries, he was the first to see bacteria, blood cells, and human sperm, which he ascertained caused fertilization to occur when it penetrated the egg. Magnification, whether in telescopes or microscopes, continues to open windows on the unknown. Satellites beam down images of Earth that reveal deforestation, desertification, or other phenomena. The Hubble Space Telescope has shown distant galaxies that are invisible or blurred by the atmosphere when observed from Earth's surface. Near Geneva, CERN's Large Hadron Collider enables scientists to study subatomic particles. Many technologies make available sensations that would otherwise remain imperceptible. For example, underwater microphones record whale songs, and cameras lowered into seething volcanoes show glowing lava too searing to be approached in person. As these examples suggest, there is a large and growing class of objects invisible to our senses that mediating technologies can make known. In addition to the brief examples already mentioned, this chapter will consider holograms, IMAX films, satellite images, drone photography, images from the Hubble telescope, and the Mars Exploration Program.

Holograms seemed tantalizingly close to widespread adoption a generation ago. The leading scholar on holograms, Sean Johnston, sees them as a form of "pictorial spectacle that can evoke the technological sublime."[7] He has traced technical developments in holography wherever they occurred. Johnson notes that as with other spectacular technologies, their sublimity "was evanescent: the awe engendered was fleeting and demanded constant extension to maintain enthusiasm" among the public. "Like an addictive drug, ever more was needed to produce its [psychological] effect."[8] Johnston has carefully reconstructed the stages in laboratory work of a wide gamut of researchers, including counter-cultural groups who avoided mainstream funding and scientific publication. Holograms became interesting to artists, leading to exhibitions from 1968 onward, including remarkable early work in Ukraine. Often, these exhibitions emphasized dramatic effects, versatility, and wonder more than aesthetically pleasing works that mainstream art institutions were ready to admire.[9] After viewing a major hologram exhibition in 1975 the *New York Times* critic Hilton Kramer sniffed, "It is difficult to know which is the more repugnant, the abysmal level of taste or the awful solemnity that supports it."[10] Commercial applications of holograms proved more limited than investors expected, and some museums and corporate divisions devoted to the new technology closed down. What survived was the *idea* of creating three-dimensional environments entirely made of hologram imagery, perhaps most famously presented as the Holodeck (or three-dimensional visualization room) in the second Star Trek television series.[11] It remains more a potential than a fully realized intangible sublime.

Some scholars in media studies consider particular films sublime. Brady questions this. She contrasts the experience of seeing a tornado advance toward the front porch where one is sitting to viewing a similar scene in the cinema. "The filmgoer is likely to feel strong emotions and imaginatively engage with the film's content—both disaster and horror films we know can feel like real experiences of terrible things." Nevertheless, many of the environmental effects are absent. There are no powerful winds, the barometric pressure does not drop precipitously, rain does not slash across one's face, and much else is not included in the film experience. It is "contained or bounded" and lacks the "qualities of immensity, formlessness, and wildness" of an actual tornado.[12] Crucially, the filmgoer does not fear death and need not seek refuge.

The objections to IMAX films being considered sublime are similar. IMAX theaters project large, detailed images on a screen that is ten meters high. By the 1980s such theaters began to be built adjacent to popular tourist sites of the natural sublime, including Grand Canyon, Yosemite, and Niagara Falls, packaging an array of tourist experiences into tightly crafted entertainment.[13] The National Aeronautics and Space Administration (NASA) commissioned IMAX films of its space flight programs and presented them at Washington's Air and Space Museum in a specially built theater with an enormous screen and powerful speakers, providing an overwhelming "panoramic realism." Kornelia Boczkowska has examined three such films produced in the 1980s and 1990s.[14] She builds on the work of Haidee Wasson, who argued that the IMAX films provide a visualization of "the moment of our encounter with Burke's sublime."[15] This is not quite the case, however. IMAX films provide encounters that would not be possible for human beings in the flesh, such as flying like a bird or watching a rocket launch from nearby without being incinerated by the searing heat of the blast off. Yet this is not the sublime of either Burke or Kant. Sublime experience is overwhelming because it assaults all the senses at once. An IMAX film uses a very large screen, but it reproduces only sight and sound. Those who witness a rocket launch also experience the shaking of the earth and the smell of the burning fuel. Such multiple, simultaneous sensations are an essential part of perceiving a sublime object, and in this sense the IMAX film falls short. Likewise, in a comfortable movie theater there cannot be the same sense of danger that inspires fear bordering on terror when encountering a tornado or an erupting volcano. Moreover, such films do not reveal something that is otherwise inaccessible to the senses. They are substitutions for actually going to see a rocket launch or for visiting the Grand Canyon or Niagara Falls. Most IMAX films therefore are representations of the natural, technological, disastrous, or martial sublimes, though a few made in orbiting space stations do offer otherwise inaccessible experiences to a wide audience.

In contrast, satellite images usually are examples of the intangible sublime. They can reveal otherwise invisible structures and relationships on Earth. They provide a point of view that no one on Earth's surface can replicate.[16] They make visible the natural world's transformations and rhythms, such as the annual expansion and retreat of the Arctic ice cap. Since 1972 NASA's Landsat satellites have used remote sensing to photograph Earth's land and coastal regions. Using infrared sensors, they monitor such things as energy use, mining, and deforestation. These images can also detect thermal changes in areas of volcanic activity. They create new kinds of knowledge about climate change, desertification, and other phenomena that take place on a large scale. They produce images that are unlike ordinary photographs because they must be constructed from the satellite data, possibly adjusted slightly when compared with measurements made on Earth and then produced as images. Many of these depict vast landscapes in the tradition of the mathematical sublime.

Such images make possible projects such as "Digital Earth," a visualization of statistical information calculated to demonstrate the history and condition of the planet. In Andrea Ballatore's reading, Digital Earth is "a techno-scientific myth" that imagines a child exploring geography, history, and culture, in an experience of the "digital sublime." The technology was championed by Al Gore, lost federal funding under President George Bush, was acquired by Google, and later emerged as Google Earth. The project might appear to be merely descriptive, but it promotes a vision of wholeness and integration, in which human beings and the world are encompassed in a totalizing system.

Like other landscape modes of the sublime, the intangible sublime offers an encounter with vastness stretching toward infinity. It is akin to Kant's mathematical sublime, but with the difference that the mediation involved has become more complex than the simple magnification of a hand-held telescope. In many cases, it provides nothing to the eye except representations based on data, and the experience is entirely dependent upon mediating technologies. The experience is static, and it is only possible after the image is taken, or the data collected.

In contrast, the second mode of the intangible sublime is an interactive spectacle that unfolds in time. It is based on the experience of controlling a vehicle from a distance and having extended interactions with it. As these experiences are repeated, the user acquires a kinetic sense of the vehicle's capabilities and begins to identify with it. Drones provide an interesting example. The next chapter will examine military drones, where this interaction involves combat, but for civilians, drone photography offers an expansion of perception. Drones provide experiences of landscape that it would be impossible to see or in some cases to endure. Drones can show Earth as it looks to migrating birds or offer new visions of places already canonized as part of either the technological or natural sublime. Lev Grossman astutely realized that a "drone isn't just a tool: when you use it, you see and act through it—you inhabit it. It expands the reach of your body and senses."[17] Something similar can be said about other intangible sublimes. The technological extension of the senses can reach into many different situations and contexts. Civilians use drones to make photographs that expose otherwise invisible vistas. The *New Republic* reviewed *Dronescapes*, a book on drone photography, and concluded that its imagery "represents a kind of twenty-first-century sublime, with all of the beauty drones reveal and none of the terror." Admittedly, "Much of the collection is coffee-table kitsch: tiny swimmers bobbing in turquoise tropical waters; an Umbrian walled city enveloped in fog; a parking lot packed with cars that form a geometric zigzag; an entire chapter of wacky stunt wedding photos." Nevertheless, these images have a "transcendent quality" that comes from "precipitous height, rather than color or composition." Drones make it possible for people "to see themselves and their surroundings from a godlike vantage."[18]

Some uses of drone photography are clearly not sublime, however. They are used in espionage, and amateur operators surveil neighbors without any legal right or even a clear motivation. Yet drones make possible new ways of seeing as they "collect visual signals, detect heat and moisture, and record radiation, space, and sound. Like the roving, collecting 'camera eye' of the early twentieth century, the drone eye acts as a visual idiom for its time and place. But even more than the camera eye . . . [the] drone eye's view is also a threat. Every datum it takes in contributes a tiny piece to a vast project of targeted annihilation."[19] Neither military attacks nor spying nor invasions of personal privacy can be considered sublime. No technology is automatically sublime. Depending upon how it is used, the same device may enable a sublime experience or kill innocent people. The many possible uses of drones emphasize that a technology is not merely a system of machines with certain attributes; it is always part of a social world. The drone's uses are not inevitable: each is embedded in institutions and social processes that shape its uses. Every technology is an extension of human lives. Who makes it, who owns it, who uses it, and how it is interpreted all matter. A technology is not a thing-in-itself, but an evolving set of problems and possibilities.[20] The social construction of the drone has only begun, and it has many potentialities.

Like an extraterrestrial drone, the Hubble Space Telescope, directed by a large team of scientists, engineers, and other specialists, has provided spectacular images of outer space. The team developed a long-term relationship with the Hubble. Telescopes have long offered vivid examples of the mathematical sublime, and just as the discoveries of Copernicus, Kepler, and Galileo expanded the possible scope of the natural sublime, more powerful telescopes continually provide new images of outer space. In 1990 NASA launched the Hubble telescope, which is roughly the size of a school bus, into an orbit 569 kilometers above Earth (figure 5.1). It has transmitted 10 terabytes of data each year and made more than 1.3 million observations.[21] Since no atmosphere obstructed its view, the images derived from the Hubble's data are sharper and more detailed than those from any telescope on Earth. As the Hubble's historian Robert Zimmerman noted, these images have stimulated public interest in astronomy and led to increases in funding. Hubble images were "eagerly viewed by millions of people almost immediately upon release."[22] Elizabeth A. Kessler examined these photographs, particularly those produced by the Heritage Project, a team of scientists and technicians inside NASA responsible for producing and disseminating Hubble photographs. These were not merely selected; they were constructed from the data sent by the Hubble and from other sources. The group self-consciously drew upon

5.1 Hubble Space Telescope hangs 569 km above Earth's surface, where it can avoid the atmosphere and make clear photographs of objects in space. Courtesy of NASA, https://photojournal.jpl.nasa.gov/catalog/PIA18165.

the sublime tradition. They saw the imagery of outer space as an extension of Ansel Adams's photographs of the American West, and they were aware of the traditions of nineteenth-century landscape painting and photography.[23] When the Heritage Project made a series of black-and-white images of the universe, they privately called it their "Ansel Adams Gallery." His austere images of deserts and mountains in the United States had prepared them to see grand astronomical spaces. Moreover, romantic paintings had often paid great attention to cloud formations and to the obscuring effects of fog and mist, and these served as a visual analogy for images of nebulae and gas clouds in space. Interviews with the group revealed "the profound commitment of astronomers to conveying the awe they feel when observing the cosmos."[24]

One of the directors believed that in addition to the Hubble's scientific mission, its images gave ordinary people a sense of "how large the cosmos is and how small the planet Earth is," and, as Kant would expect, this could be "a little negative because it puts you in a very diminutive role."

Yet the achievement of launching a satellite with a large telescope and sending these images back to Earth also meant that "the dominant emotion should be a sense of awe that we can understand this stuff at all."[25] The public comments left on a website where the images are displayed suggest that many did experience what Kant would call the mathematical sublime. One viewer found the photographs "more than just a sightseeing trip through the universe. These pictures shake the foundations of everything I have ever believed in and leave me reeling, breathless, and stunned at the wonders that exist beyond the scope of our imagination." The only consolation was "knowing the closest I will ever get to these wonders is through my computer screen."[26]

These images are not the straightforward representations that they appear to be. Kessler examined different photographs of the Whirlpool Galaxy, for example. She found that when astronomers used one in a scholarly article for *Astronomer's Journal*, they "optimized" details for their purposes. By dampening the "dramatic variation in brightness" they made the "regions of the galaxy exhibit a similar degree of structure, even the core where the brightness of the concentrated stars usually obscures the morphology." This result did not satisfy the Heritage team, however. For broad public consumption, they combined that image with another one, with the intention that the nucleus of the galaxy be, as one of them explained, "much brighter than the arms." Even so, the result "is unrealistic in the sense that the nucleus is *really* a lot brighter than the spiral arms." But this was not all. The Hubble image only showed the center of the Whirlpool Galaxy, so the Heritage Project team "also incorporated an image from a ground-based telescope" to fill out the periphery.[27] The team combined three images to create what appears to be a single, realistic view. Considerable artifice was employed to create a photograph that resembles what a human being in outer space might see looking through a telescope. The images that the public enjoys turn out to be constructions, not mirrors. Moreover, the aesthetic employed comes from art, not science. Kessler compares the Hubble images to Thomas Moran's landscape paintings of the Grand Canyon, which made little attempt to depict scenery realistically, but rather combined various elements of the landscape into a composite image. "Neither Moran nor those who crafted the Hubble images felt compelled to preserve precise relationships between different elements in nature."[28] Both created mediated visions.

The experience of driving the Mars rovers provides an excellent example of the interactive, immersive mode of the intangible sublime. Teams on Earth landed sturdy, compact rovers on Mars and then used them to explore the planet. NASA designed the rovers to study the evolution of Mars and its potential to host life.[29] These teams had an interactive relationship to the rover as they moved with it through deserts and mountains. They instructed rovers about what photographs to take and what data to collect on the planet's climate and geology. After landing in different locations in 2004, the rovers Spirit and Opportunity almost immediately revealed that there had once been streams of water flowing on Mars. Expected to work for 90 to 100 Martian days and to explore about one mile of the surface, Opportunity continued to function for more than fourteen years. It examined a wide range of landscapes, including a crater whose rocks were four billion years old.[30] In February 2021 NASA landed another rover on Mars, Perseverance, which weighs a bit more than one ton. It touched down in the Jezero Crater, chosen because it had once formed a lake and therefore might preserve signs of former lifeforms. Like the Hubble Space Telescope, NASA's rovers provide a stream of information and photographs that are available to the public at no cost. Perseverance began sending photographs back to Earth as soon as it landed, including a panorama of its location that NASA stitched together from seventy-nine individual images, and a picture of the Jezero Delta, which was two kilometers away (figure 5.2). Perseverance also had a Twitter account, where it posted every day. It announced on March 9, 2021, that it was "flexing my robotic arm and doing some more checkouts of my tools" and announced that for several more weeks it would be completing "health checks." By the next day these comments had been retweeted more than 5,500 times. The Twitter account and the language used there made Perseverance seem more human. Like a ventriloquist with a dummy, NASA gave its robot a voice to interpret its mission to the public, and less than a day later sixty thousand people had "liked" the tweet and many had tweeted back.

The men and women who drive the rovers on Mars have an interactive experience. Ed Dolan, after working for years on other Mars projects,

5.2 The ancient river delta on Mars at the edge of the Jezero Crater may contain evidence of life. Photographed by the Perseverance rover on March 5, 2021, from 2.3 km away. Courtesy of NASA, https://photojournal.jpl.nasa.gov/beta/catalog/PIA24485.

found that controlling a rover was not at all like monitoring satellites. "I spent my whole career looking down from above. And now we're down on the ground. It's a very different experience from what I had before."[31] The satellite was panoramic, sending back photographs of large swathes of the Martian surface. It moved in orbit and did not require as much detailed daily planning. But each Mars rover's movements and actions, such as sampling soil or rocks, had to be determined based on the exact location where it had stopped. Because of the distance to Mars, there is always a time delay of at least seven minutes, and this meant that a rover cannot not be driven continuously but rather must make many short advances. At every stop, the driver needs to study the terrain, adjust the trajectory, and estimate how far the rover could safely move. A team of experts shared their assessments of the situation. An accident could immobilize or even topple the rover. Too little exposure to sunlight and its batteries would not recharge. Rough terrain could harm the vehicle; sandy areas could prove a trap. In their efforts to know exactly what a rover can do, the drivers became thoroughly identified with their machines. They found that *"seeing like a rover"* required "learning to see like a member of the Rover team in the context of a particular (and peculiar) social organization, and to account for robotic and human relations in particularly intimate, totem-like terms."[32]

During pauses in each rover's movement at the end of the Martian day, they met to decide "what instruments the MER will use, what 'observations' it will make, where it will go and for how long. It determines when and where the rover will drive or be idle (to recharge its solar batteries, referred to as a 'siesta.')"[33] Before the beginning of each Martian day (or "sol"), the team must assess the data from the day before, look at an oversubscribed list of possible actions from scientists, decide on a plan for the next drive, lay out the exact sequence of actions to be performed, and then transmit these orders to Mars.[34] Because it is not safe to drive a rover during darkness, the team must work, sleep, and eat according to the rhythm of Martian days. They do not see a planet from an orbiting satellite. They are engrossed in dramatic movement on the ground. One team, in interviews, called their activity exploration and compared themselves to Columbus and to Lewis and Clark. They did not see their work as a job but as the apex of their professional careers, and often they came in early to see the new images and data as soon as they became available from the day before. Their mission was indeterminate in length, depending on how long they could keep their rover functioning. It could end after only a few months, or it might last for years. They continually developed plans, deciding the rover's route, what it should sample, and what deserved closer inspection. Possible discoveries remained open-ended and changed in response to what they found. Yet, the mission could suddenly end at any time, and the "idea that the rover might get stuck is experienced as death."[35]

If NASA press releases and Twitter postings encouraged the public to understand the rovers in anthropomorphic terms, the Opportunity team identified with its rover, especially the drivers. To take photographs using its equipment, they learned to identify various parts of the body with the parts of the rover, in order to understand the right position to assume when taking a photograph. To do the job well, the driver needed to develop an "embodied imagination," which expressed not only the relationship between driver and rover, but the collective relation between a large team operating the mission and the rover.[36] One team member explained, "It's been some kind of weird, man-machine bond [laughs]. It's become an extension of each one of us, our eyes or our hands, our feet. . . . I guess in a way, it's through them that we are tasting, tasting the rock. It's . . . kind of, it has morphed into us, or we've morphed into it." Gradually, they learn what exactly the rover's capabilities are, and, as another team member put it, they "inhabit it" or project themselves "into the rover. It's just an amazing capability of the human mind. . . . that you can sort of retool yourself."[37] Such identification can also be found among oceanographers directing undersea robots.[38] Compared to NASA's team, the public cannot develop the same intense embodiment with the rovers, but they can identify with those who operate them, and many take great interest in their explorations. The public experience is that of a landscape, while the team members experience an immersive spectacle.

Yale University physicist Priyamvada Natarajan noted the power of the intangible sublime to transform public perception of the universe: "Astronomical images have agency: They shift our view of the cosmos from the imagined to the reasoned and the real. This happened more than 400 years ago, when Galileo presented six watercolor images of the moon as he saw it through his repurposed spyglass—which we now know as a telescope. It happened about 200 years later, when Louis Daguerre made the first astronomical photograph—also of the moon. And it happened in 1990, with the first image of the infant universe, taken by a NASA satellite, and in 2017, with the Hubble Space Telescope's images of clusters of galaxies."[39] Natarajan hoped that the five petabytes of data collected by radio telescopes from around the world, which were collectively called the Event Horizon Telescope, would reveal the edges of one of the largest black holes in the universe. Such an image might further confirm Einstein's theories, and it had "the potential to redefine the cosmos once again, and prompt wonder and curiosity about our place in it." A few days later, a striking image of a black hole, the first-ever such photograph, was ceremoniously presented to the public and became front page news.

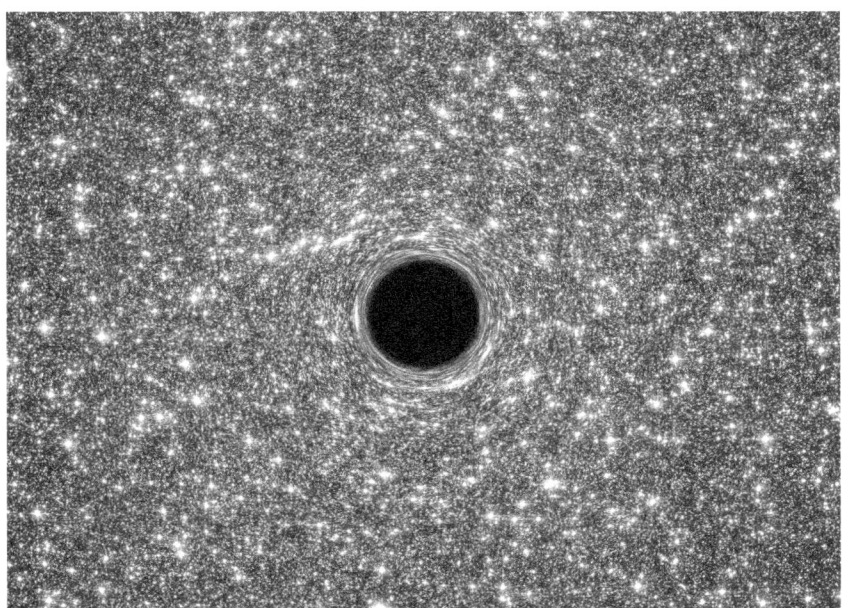

5.3 Supermassive black hole, located in the middle of the ultradense galaxy M60-UCD1. Its weight approximates the collective weight of twenty-one million suns. Its intense gravitational field warps the light of the background stars to form ring-like streaks along the dark edges of the black hole's event horizon. Photograph by Hubble Space Telescope. Courtesy of NASA, https://images.nasa.gov/details-GSFC_20171208_Archive_e000984.

Einstein's equations had led to the conclusion that when a great deal of matter or energy was concentrated in one location, a black hole might be formed where space and time collapse, "trapping light and matter in perpetuity."[40] The Hubble Space Telescope has also made images of black holes, including one in the middle of an ultradense galaxy (figure 5.3).

Scientists know that the Event Horizon Telescope's image of a black hole—never before captured—increased public understanding of (and willingness to fund) their research. It was no ordinary photograph. Eight radio telescopes on four continents had to focus on the same distant location and, using atomic clocks to coordinate their observations, gather data at precisely the same instant. Mere synchronization was not enough, however, as the data had to be gathered when weather conditions were perfect at all eight observatories "in Europe, North America, South

America, Hawaii and the South Pole."[41] This coincidence finally occurred in April 2017. It then required two years to analyze the data at the Max Planck Institute for Radio Astronomy and the MIT Haystack Observatory. Only at the end of this laborious process could the scientists produce the iconic image of what looked a bit like a vast, round blank, surrounded by clouds of photons and particles that from Earth's perspective swirled to the inner edges of the circle and disappeared into it.

The production and display of this image exemplify how esoteric scientific projects can be presented and explained to the public. If the scientific reasoning behind black holes is too abstruse for all but physicists, the image itself is striking, stirring wonder at the strangeness of the universe. It is even more impressive when one learns that it is 55 million light years away and that the gravitational field of this black hole is 6.5 "billion times more massive than the sun," so powerful that nothing can escape it, not even light. This is a scale of power and distance far beyond what human beings can perceive. Like Kant, we must use reason to understand it better. In just one hour light travels 669,600,000 miles, which is already hard to grasp. An object one million light years away is 8.76 billion times further than one light hour, and yet that is less than 2 percent of the distance from Earth to the black hole photographed. To describe such phenomena in a way comprehensible to laypeople requires both a mastery of the science, uncommon writing ability, and a knack for finding appropriate metaphors and figures of speech. Fortunately, modern science has spawned a remarkable outpouring of popular writing, such as Stephen Hawking's *A Short History of Time*.

Unlike the traditional sublime, which dealt with objects the senses can apprehend (whether natural or technological or a combination of the two), many objects of contemporary biology, physics, and astronomy cannot be sensed. No one can see, hear, smell, taste, or touch a fragment of DNA, a neutrino, a quark, a Higgs boson, the dimensions of string theory, or a black hole. Because we cannot sense them directly, these objects must be represented through mathematics, images, graphs, diagrams, and descriptions. To realize their sublimity requires the assistance of scientists who are also skilled writers, such as Steven Weinberg or Richard Feynman.[42] The technological sublime refers to tangible objects encountered directly, but the many objects of the mediated sublime can only be apprehended through technologies that teams of researchers use to extract data and construct images, elegant drawings, and explanations. This sublime experience first is mediated by a technology such as an orbiting telescope that sends data back to Earth, and then by scientists who take this information and transform it into images and interpretive prose. The public consider such experiences sublime in keeping with the original meaning that Longinus had for the sublime, that it is produced by persuasive language that lifts an audience out of themselves and enthralls them. For scientists to make their work accessible to the layperson, they must become rhetoricians who can explain the microscopic building blocks of life, subatomic particles, and distant galaxies.

As the distance between subject and object widens, phenomena become invisible to the senses. Kant could look through a telescope and see the same things as the scientists of his day. But subsequently, the scientific sublime became a form of double mediation, first by using mediating technologies that gather data that is only intelligible to experts; second by writers explaining discoveries to a wider public. This also includes explanations of temporality. Some experiments conducted on subatomic particles take place in far less than a second; some astronomical events have unfolded over billions of years. The human sense of time, based on days, years, and life expectancy, is challenged to grasp events on either the microcosmic or macrocosmic scale. This intangibility is partially overcome when a team operates in human time and creates an intimate relationship with the distant vehicle as they learn to control it.

Considering all the forms of the sublime examined so far, only the intangible sublime presents no immediate danger. In the disastrous sublime, the earthquake, hurricane, or flood is not only palpable but also a mortal threat. In the martial sublime, the dangers are so overwhelming that they can impair the ability to stand back and reflect on the experience, and they border on the traumatic. The intangible sublime presents no immediate danger as one contemplates astonishing subatomic particles, black holes, and other spectacles. But for all its wonders, this sublime carries with it an intimidating realization: that humanity is only a tiny, vulnerable speck of life in a vast and quite possibly indifferent universe. There may be no present danger, but there is no assurance that humanity will survive. The surface of the Moon or Mars, both pockmarked with craters, are reminders that massive meteors have struck Earth in the past and may well do so again, causing mass extinctions. Should humanity escape that fate, we know that our sun will not burn forever. And, in some distant future, our solar system could be destabilized by the powerful gravity of a black hole, where it will disappear and be crushed. The intangible sublime of astronomy may present no immediate danger, but given the perspective on the universe it provides, there appears to be no escape from eventual annihilation.

6

DIGITAL?

During the last decades of the twentieth century discussions of digital technologies sounded utopian, as is common when new technologies emerge.[1] Computers were variously expected to undermine patriarchy, bridge cultural barriers, enable a frictionless global economy, democratize access to knowledge, energize grassroots activism, empower women and minorities, and become an engine of re-enchantment.[2] Since then, fears have tempered these hopes due to the spread of disinformation, corporate domination of the Internet, mobbing and hate speech on social media, terrorist websites, government surveillance, identity theft, and cybercrime.[3] Despite these problems, digital technologies make possible new experiences. Are any of them sublime?

Burke considered vastness a sublime attribute, but he also appreciated the apparently endless "scale of existence" that stretched down to ever smaller entities: "we become amazed and confounded at the wonders of minuteness; nor can we distinguish in its effect this extreme of littleness from the vast itself. For division must be infinite as well as addition."[4] One can be amazed by miniaturization, as the information packed on a silicon chip has doubled and redoubled every few years for decades. Kant's theory can also be linked to these achievements. Recall that, as John Goldthwait summarized, for Kant, "The feeling of the sublime is the feeling of our own inner powers, which can outreach in thought the external objects that overwhelm our senses."[5] The mind can conceive a vastness or a minuteness greater than the senses can perceive when contemplating digital hardware. Digital hardware can excite wonder when one understands its complexity, speed, global reach, and the enormous scale of its data storage centers. As one philosopher argues, "In a world in which the computer has become the dominant technology, everything—genes, books, organizations—becomes a relational database. Databases transform everything into a collection of (re)combinatory elements. In this sense databases have become the dominant 'cultural form' of our age."[6]

Images of server farms employ visual strategies pioneered 130 years ago in photographs of vast textile mills and factories, where long rows of machines stretched to a vanishing point.[7] In 2012, when Google decided to emphasize its infrastructure in a public relations campaign, this strategy was used again. Its images showed row upon row of servers, symmetrically organized, with no human beings in sight, as if the system ran itself. Another commonality with the technological sublime is the use of impressive measurements and figures. Celebrations of skyscrapers, factories, and hydroelectric power stations employed a kind of romance of numbers.[8] For example, in 1931 promoters of New York's new Empire State Building had a parade of statistics to impress the public. It was the world's tallest building and was completed ahead of schedule. It contained 67,000 tons of steel, more than had ever been used in any building before, as well as 200,000 cubic feet of Indiana limestone and 10 million bricks. Its interior enclosed 37 million cubic feet of space.[9] In 2018 IBM employed a similar strategy to celebrate Summit, the "world's fastest computer" (figure 6.1). Summit weighed as much as a large commercial jet. It needed 4,000 gallons of water per minute to keep it cool, and it could make 200 quadrillion calculations per second. By comparison, if "every person on Earth completed one calculation per second," it would take them ten months to do the same.[10] Two years later, the Japanese put into operation the Fugaku computer, which is twice as fast. Photographs of such machines and of server farms depict landscapes of immense miniaturization, coupled with statistics about speed of calculation and the scale of the data being stored.[11]

But in contrast to the dams, canals, and skyscrapers of the technological sublime, such landscapes are not central to the public's sense of the

6.1 The IBM Summit Computer at Oak Ridge National Laboratory was the world's fastest from 2018 until 2020. Courtesy of Oak Ridge National Laboratory, https://www.ornl.gov/sites/default/files/2018-P03963.jpg.

digital sublime. Instead, the focus of the digital sublime is on software, though not the many programs designed to perform mundane tasks such as payroll management. They are no more sublime than the typical office tower. But some software mesmerizes an audience. Eugénie Shinkle notes that "a feature of the digital sublime is the absence of a consistent and uniform boundary between the self and the machine," and this boundary commonly seems to dissolve when playing online games or exploring virtual reality (VR). Full verisimilitude is not necessary. For example, early users of flight simulator programs found them acceptable, even though the landscapes traversed were rudimentary sketches.[12]

The term "virtual reality" emerged in the late 1980s. Some argue that its precursors include medieval cathedral windows and the eighteenth-century phantasmagoria, which used projections from magic lanterns and sound effects to create a multimedia experience.[13] Miles Orvell emphasizes the stereoscope as a precursor to VR. Invented in 1838 and adapted to photography in 1849, it created a sense of three-dimensionality.[14] Stereoscopes were widespread in Victorian homes, which acquired sets of cards that depicted many sublime sites, including European capitals, the Alps, Niagara Falls, Civil War battlefields, railroads, national parks, and the first skyscrapers. Other predecessors of virtual reality include 3D films, in which the audience wore stereoscopic glasses, and Cinerama, which wrapped a curved screen around the audience. Each of these increased the sense of immersion, as "the horizon" between user and apparatus seemed to evaporate so that "a mediated experience feels immediate." This horizon is constantly "being redrawn between human and machine," as technical capabilities improve and aesthetics change.[15]

VR emerged in 1960, when Morton Heilig invented the first individualized VR display. Worn over the eyes, it provided stereoscopic images and stereo sounds, but it was not interactive. Eight years later at Harvard, Ivan Sunderland developed a head-mounted display that streamed stereographic computer graphics and adjusted the images as it tracked a user's head and eye movements. However, the components were not well integrated.[16] Progress was slow in the 1960s and 1970s because, as Carl Machover recalled, "Once you begin to build 3D software for design purposes, you run into equipment limitations: software is pretty expensive, it's quite hard to learn, and [it] required significant resources from a computing standpoint."[17] Even in the 1980s it remained "prohibitively expensive" and "the computer power necessary to portray a visually rich world did not exist."[18]

There was funding in government laboratories, however, notably at NASA's Ames Research Center, which experimented with a head-mounted stereoscopic device and specially designed gloves that enabled the illusion of interaction with the virtual world. By 1988 it could offer a goggle-wearing user "a multisensory, interactive display environment in which a user can virtually explore a 360-degree synthesized or remotely sensed environment and can viscerally interact with its components."[19] The

Ames Research Center later developed a full-body garment that enhanced the experience, and it developed software that enabled others to write VR applications.

As the cost of computers and memory chips fell, private companies also began to design interactive virtual environments. Yet as late as 2010 few people had experienced VR. As Machover explained, "This idea of being tethered to equipment through a helmet or something is intellectually acceptable, but practically, it's not very acceptable. That's one of the reasons why you get into some of the virtual reality rooms that surround you, so you can go in them with the minimum of encumbrances. Even there you put on special glasses and things of that type. Obviously if you had some dire need for it, if you were an experimenter, you'd put up with those inadequacies, but a casual observer is not going to put up with a bunch of stuff. You have to make it very easy for them to use."[20] One of the problems was integrating haptics with sight and sound. To work well, a glove or a body suit required extensive wiring and many sensors linked to a computer, without becoming cumbersome. Sensations were most convincing if the suit was a snug fit, but one size hardly fit everyone. Given these problems, early VR remained largely visual and aural, and even in 2020 the most common technologies were gloves and headsets of various types.

If VR advanced slowly, simulators for training aircraft pilots and steersmen for large ships emerged during World War II. In 1943 MIT signed a contract for the Whirlwind Project, to develop a simulator that could mimic different aircraft, rather than build separate simulators for each sort of plane. Because this kind of simulator required as many as one hundred thousand operations per second, it demanded an entirely new kind of computer. That led to the development of core memory, which was faster and more reliable than the alternatives. The work took eight years at an astronomical cost, but the new memory system proved valuable to the military and became the basis for the IBM AN/FSQ-7 computers at the heart of the SAGE defense system. That system cost $8 billion, but the technologies developed, including printed circuits, core memories, and mass storage devices, would prove crucial to computer development generally.[21]

In the early 1990s *Computing Systems in Engineering* published a study of the "kinematic and dynamic constraints" that faced researchers seeking to program effective simulators. It found that human beings bring to the perception of virtual worlds "considerable *a-priori* knowledge about the possible structure of the world," and much of their "sense of physical reality is a consequence of internal processing rather than being developed only from the immediate sensory information we receive."[22] To create the illusion of another world, designers had to understand what human beings assume about the structure of reality. "The fact that actors in virtual environments interact with objects and the environment by hand, head, and eye movements, tightly restricts the subjective scaling of the space" and any "mismatch in the gains or position measurement offsets will degrade the performance by introducing unnatural visual-motor and visual-vestibular correlations." The difficulty of coding continual adjustments increases when creating the illusion that a person is walking through a virtual world. Virtual space must continually change in response to movements, and the changes need to be instantaneous. Any time-lag will "interfere with complete visual-motor adaptation."[23] Complicating matters further, experience with simulations showed designers that they had to use "information distortion" if they wanted "to produce illusions of greater freedom of movement."[24] Solving this tracking problem slowed down VR's development, because it demanded enormous computing power and complex programming to create convincing interactive illusions.

Long before VR could be perfected, the novelist William Gibson popularized a different idea of cyberspace, which he described as "a three-dimensional, immersive experience," in contrast to later descriptions of cyberspace as "a transglobal network."[25] Gibson's characters merged into the digital world through implants, and his cyberspace imaginary was not a network like the Internet, but a space experienced when hooked up through the nervous system, or "jacked in." Gibson's hero, Case, feels most intensely alive when jacked in, able to enjoy a bodiless life in cyberspace. Those who entered "the matrix" were "surrounded by bright geometries presenting the corporate data." Furthermore: "Towers and fields of it ranged in the colorless nonspace of the simulation matrix. The electronic consensus-hallucination that facilitates the handing and exchange of massive quantities of data" was surrounded by walls of electronic protection that Gibson called "ice."[26] In his 1984 novel *Neuromancer*, he described cyberspace as a "consensual hallucination experienced daily by billions of legitimate operators, in every nation, by children being taught mathematical concepts. . . . A graphic representation of data abstracted from the banks of every computer in the human system. Unthinkable complexity. Lines of light ranged in the nonspace of the mind, clusters and constellations of data. Like city lights, receding."[27]

This is not Kant's sublime, though there are affinities with the mathematical sublime's confrontation with immensity. Gibson's sublime has "unthinkable complexity," visualized as "lines of light" perceived only in "the non-space of the mind." Gibson uses images of electrified cityscapes to express this complexity. But where the night city exists in physical space, cyberspace does not. It is a graphic visualization of abstractions that remain ungraspable, "like city lights receding." The mathematical sublime demonstrated the power of human reason when contemplating the absolutely great in nature. Its ultimate reference point was the infinity of the nonhuman world. The technological sublime arose when people encountered the seemingly infinite complexity of constructed domains. For example, both corporate executives in skyscraper offices and tourists on the observation deck found that aerial views provided a sense of mastery over the vast scene, including the tiny people crawling along far below. That sublime was built from materials wrested from nature; but it was still a solid, physical world.[28] Gibson's cyberspace, in contrast, was a "consensual hallucination" that imitated some aspects of the material world, but its objects had no physical correlative. Cyberspace promised to overcome the body's limitations and to link billions of minds in a nonspace where consensual hallucinations became the norm. The movement from mathematical sublime to technological sublime to cyberspace traces the shift from a transcendental nature toward increasingly mediated experience. The technological sublime based its exaltation on vast engineered constructions and the complexity of the built environment. Gibson's digital sublime left the physical world behind and posited unembodied existence. In *Neuromancer*, the character Dixie Flatline exists only in cyberspace.

The cyberspace Gibson imagined in the 1980s has not (yet?) been realized. Nor did virtual reality then become widespread. Rather, the Internet emerged from a U.S. Department of Defense project.[29] In the early 1990s the invention of the web browser made it possible for anyone to surf the emerging Internet and to create a home page. Skilled teams slowly developed VR at great expense. In contrast, the Internet decentralized development to a wide public, and online content exploded. This was not the cyberspace that Gibson imagined. One did not jack into the Internet, linking the nervous system to an online world, temporarily leaving the body behind. One did not go behind the images to see visual representations of computer code. And one did not share experiences directly with other minds. People sitting in front of computer screens in different locations were in different contexts and did not necessarily share experiences even when viewing the same websites. Their screens contained different advertisements, based on their location and browsing history, and they were interrupted by different emails and alerts. Each viewer was drawn from one site to the next, making a chain of connections unlike the sequence viewed by anyone else. The on-screen "landscapes" were as heterogeneous as people could make them.

The Internet proved to be the source of ambiguous information. Frequently, there was a "cognitive rupture in which the relation of signified and signifier breaks down" due to "an excess of material." As Annie Dorsen deftly explains, unstable meanings on the Internet take two forms.[30] The first occurs when a search produces thousands of contradictory hits, and the user is "bombarded by an excess of signifiers" creating a sensory overload. This creates "a massive underdetermination that melts all oppositions or distinctions into a perceptional stream." So much bombards the user simultaneously that it is impossible to assimilate and analyze all the signifiers (words, images, and sounds). The second possibility is that the Internet provides an excess of signifieds, which creates a massive overdetermination, "in which one can read so much into a given image or word that it becomes overloaded, a black hole of potential meanings."[31] One reaches a state that Thomas Weiskel called "absolute metaphor," in which anything might mean anything else, or nothing at all. Whether underdetermined by an excess of signifiers or overdetermined by an avalanche of signifieds, the result is "death by plenitude."

Rampant deceit and dishonesty further vitiate Internet communication, which spreads malware, half-truths, and disinformation. Users must be wary of phishing, identity theft, unreliable information, and manipulated images. Hacking has become a profitable underworld business, and national intelligence services operate as "legal hackers." Surfing the Internet is like walking on a crowded street with many pickpockets and spies but few police. It has become a series of wary encounters with unsafe websites, corrupted files, self-serving content, criminality, fragmentary news, and interactive disturbance. The Internet is not Gibson's "shared hallucination" but a nonspace where meanings degrade, fake news proliferates, and snippets of information go viral, while reliable information is sliced into fragments. Sianne Ngai has combined the words "stupid" and "sublime" to form "stuplimity" to describe the emotional state that Internet usage induces. Its repetitions and variations overwhelm and exhaust the viewer, who is simultaneously bored, astonished, and desensitized, leading to the death of meaning by plenitude. The Internet provides not a unified experience but a divided one; not a focused experience that engages many senses but a diffuse one involving just a few; not a powerful or awe-inspiring experience of immensity but a scattershot of trivia, advertisements, blogs, and emails, with valuable content intermixed. Hardly the description of a sublime experience.

Gaming is another matter. Online landscapes first emerged in Britain as multi-user dungeons, or MUDs. They "provided a first look at what persistent on-line worlds could be, allowing players to wander and act with a high level of freedom" in imaginary landscapes.[32] These evolved to become multiplayer virtual worlds. By the early 1990s there were thousands of MUDs, with themes from story books, science fiction, and popular culture. As computing power increased, the games also acquired more narrative elements.[33] Pac Man occurred on an abstract playfield, and there was no story. But as processing speed and graphic representation improved, games presented a world seen from the perspective of a human body, and movement through that world demanded narrative. By 1987 a game could be a spin-off from the George Lucas *Star Wars* films, which "let players chat, spend money, go on treasure hunts, and run businesses."[34] Developers lost some control as more players got involved. This period's legacy included "avatars," or the on-screen representations of players.

As games and their narratives matured, they offered more complex worlds to explore and new opponents to battle. In a few, like *Ultima*

Online, fighting was optional and limited to certain regions. Some players spent their time in online pubs or fishing on digital docks, gradually bonding with others and forming groups. A few players began to socialize offline as well.[35] But many popular games emphasized competition and combat. By 2002, at peak times *Counter-Strike* attracted 90,000 simultaneous players, and the action tended to be "bloody, focused, and fast." Collectively, 1.7 million gamers logged 2.4 billion minutes every month on *Counter-Strike*.[36] *EverQuest* became an even more popular game later that year. It did not permit players to kill one another but focused on slaying monsters. Players were encouraged to work together, and the "game was based on player cooperation."[37] In online worlds human beings seemed to merge with their machines via online avatars. But these experiences differ substantially from the sublime of either Kant or Burke. Their sublime was not about battle or struggle. It was not a narrative, certainly not a quest. It was not about winning (or losing), nor did it require killing enemies or slaying monsters.

The illusions of gaming are seductive, and players often feel that there is no longer "a consistent and uniform boundary between the self and the machine."[38] As Ted Friedman notes, "Computer games allow players to lose themselves in imaginary worlds, with incredible flexibility to mold characters and environment." Moreover, "these worlds are in constant flux" providing ever more possibilities.[39] Yet, there are limits to this flexibility. As Robert John Baron argued in an analysis of *World of Warcraft*, "the digital design of this virtual environment structures its users' interactions with the game world itself and with one another."[40] He emphasized "the programming code and guiding logics that underlie virtual environment experiences." Players have agency, but for a game to work, it must lead players "to form a cohesive virtual community" that shares sensations, attitudes, and ideas. The *World of Warcraft* game "places players in a rhetorical context and supplies a body of in-game experiences, symbols, and practices that lead users to see one another as virtually consubstantial." These experiences lead to intense, long-term engagement with a medieval virtual world populated by trolls, elves, dwarves, orcs, and dragons. This world is riven by conflict, and it demands that players take sides and adopt a clearly defined role with skills that make each player the member of a guild. This architecture embodies "the subtle ways in which power and direction can be exercised within virtual spaces. Virtual world designers and producers do not need to expressly tell the users of a virtual world what they can and cannot do."[41] As Peter Krapp concluded, "Interaction with gaming interfaces makes human control only one facet of a tightly constrained, and, indeed, controlled environment that sometimes threatens to reduce the human in the loop from a true controller to a mere 'ebkac'—an error between keyboard and chair."[42]

Games such as *Sim City* and *Civilization* are absorbing in a different way. Players have the omniscient role of creating a skyscraper, a city, or an entire civilization. They must govern, plan, build, and police, and these many responsibilities result in identification not with a particular role but with the program itself. As Friedman concluded, "Your perspective—the eyes through which you learn to see the game—is not that of any character or set of characters. . . . The style in which you learn to think doesn't correspond to the way any person usually makes sense of the world. Rather, the pleasures of a simulation game come from inhabiting an unfamiliar mental state: from learning to think like a computer."[43] This perspective provides what *seems* to be a new form of sublime experience. But the program shapes this awareness and encourages "the strange sense of self-dissolution" that players experience. They often play for hours and lose track of time in "an almost-meditative state, in which you aren't just interacting with the computer, but melding with it." This feedback loop is the essence of what passes for the digital sublime, as simulation games "aestheticize our cybernetic connection to technology."[44] This aesthetic is malleable. A game can express any ideology, including socialism, welfare capitalism, liberalism, anarchism, or religious fundamentalism. When a game fits a player's underlying world view, it can seem to mirror the construction of reality. The malleability of games encourages designers to tweak them to maximize the pleasures of gaming even at the cost of verisimilitude. As David Myers has noted, "Many representations in early digital simulation games—including representations of spatial and temporal relationships—became less realistic as they were revised and edited during subsequent, recursive play."[45] However, realism remained important, especially after circa 2010 when sport simulations became hugely popular, filling arenas with audiences to watch champion gamers compete.

While simulation games attracted millions of players, they also proved to have military applications. War games can be traced back at least to late eighteenth-century Germany.[46] Board-based war games gradually moved from tabletops to computers, notably in games designed by James F. Dunnigan and in research projects at Stanford, or similar work done at MIT and other institutions with defense contracts.[47] In the process, turn taking (in board games) gave way to simultaneous moves in real time. Already, during the Vietnam War, the Defense Department began to work toward what it called an electronic battlefield. As General William C. Westmoreland declared in 1969, "On the battlefield of the future, enemy forces will be located, tracked, and targeted almost instantaneously through the use of data links, computer assisted intelligence evaluation, and automated fire control."[48] However, it would take decades to develop and implement such capabilities.

In 1978 Atari made an enhanced version of its "Battlezone" game for the Defense Department's Defense Advanced Research Projects Agency (DARPA). Digital flight simulators and other programs soon followed. These focused on helping a pilot or tank driver acquire the skills needed to operate one vehicle. In the early 1980s the American military decided to develop expertise in coordinated actions by "a large number of combat forces working together to achieve an integrated battle outcome, especially where the battle involved a mix of weapon systems, organizations, and perhaps nationalities, which had to function together on the battlefield against a hostile, sentient opponent." Using this simulation network, called SIMNET, teams could practice "collective skills, as distinguished from the mastery of the operation of one's individual platform."[49] SIMNET could link hundreds of simulators into a single system. At meetings between industry and the military these protocols quickly "evolved into the Distributed Interactive Simulation (DIS) Standard Protocols,"[50] which also had nonmilitary applications.[51] The IEEE approved them in 1993.[52] Before then, SIMNET training had proved its efficacy in the 1991 Iraq War.[53] The Defense Department was particularly proud of one battle, in which a tank troop wiped out an Iraqi Republican Guard armored brigade without suffering any casualties. The entire battle lasted only twenty-three minutes. The U.S. Army combined SIMNET data, satellite images, and interviews with the tank commanders, to reconstruct the battle in a high-resolution 3D format.[54] It was used to train tank drivers at Fort Knox. Such simulations based on Sun workstations and networked PowerMacs proved quite effective.

During the 1990s, warfare also became a visualized event for remotely engaged soldiers in "the Synthetic Theater of War (STOW). Software was developed to cover everything from the flyout characteristics of missiles to the shading and texturing of surfaces in various lighting and weather conditions."[55] The military modified the commercial game *Doom II* to become *Marine Doom* (1997), and such developments culminated in *America's Army*, a game funded by the U.S. military and released at no charge to the public on July 4, 2002.[56] Designed both to draw volunteers to the military and to educate the public, by 2009 *America's Army* had been downloaded 242 million times and many additional software programs had been added. The American military began to elide the distinction between virtual and actual, and by 2008 the armed forces were using at least twenty-three different games to train troops.[57] Actual warfare could almost feel like a game.

Drones also became central to high-tech warfare, even as traditional battle lines largely disappeared. Bases in the Middle East and Africa and on naval vessels support armed drones that hover in constant readiness to strike. Military surveillance identifies targets, often based on data harvested from social media and telephone use. If military lawyers give clearance, drone operators at a distant air base execute an attack. In contrast, the immersive martial sublime is only possible for soldiers in physical combat, where the danger is real, and the action engages all the senses. Drone warfare poses no such danger to the pilots, but it produces psychological stress. The drone cameras transform targets from abstract locations to particular buildings and identifiable local inhabitants. There is an "uncanny spatial relationship between drone operators and their quarry," whom they may observe for days, becoming familiar with the daily life of the places under surveillance.[58] This intimacy is asymmetrical, as the people targeted know nothing of the operator. The drone pilot's sensory experience is limited by a two-dimensional screen, but an attack's consequences are more visible than in conventional warfare. Firing a weapon thousands of miles away is noiseless, the ground does not shake, and one has no tactile sense of the battle site. Yet the destruction is more

personalized than when dropping bombs from a plane. (This psychological engagement has been explored in the drama *Grounded* and the film
Good Kill.[59]) As one actual drone pilot turned whistleblower declared,
"When you are exposed to it over and over again, it becomes like a small
video, embedded in your head, forever on repeat, causing psychological
pain and suffering" and "haunting memories."[60] Not the sublime, but a
nightmare. Furthermore, there is evidence that drone warfare is less precise than the military claims. In Pakistan and Afghanistan more bystand-
ers than Taliban have been killed, and much of the civilian population
hates the drones. Their use turned many against American intervention.[61]

Just as the military has woven a network for its needs, the corporations
that control much of the Internet's traffic have moved toward a totalizing
environment, an "Internet of things." Everyday life is being permeated
with feedback loops, including digital clothing, watches, and armbands,
as well as surveillance devices that track movements, and monitor sleep,
heartbeats, and exercise regimes. Social media are not neutral channels
of communication, for they stimulate feedback, reinforce habits, note
repetitive actions, and encourage new proclivities, while doing little to
encourage the contemplation that the sublime would require. As the
leading computer security expert Donn Parker noted, "The commercial
world is going to win in the long run. [They are] Putting these cheap RFID
chips woven into clothing now, and these chip transducers are going to
be inserted into every product we have, so that every single product can
be electronically tracked."[62]

As the Internet is woven into personal possessions, the life world
becomes a commercial landscape. Amazon has developed software that
recognizes consumers by the shape of their hands and might obviate the
need to sign in or have a credit card handy. Can the sublime be found in
a virtual mall, where corporations control the flows of information, imagery, and money, and continually expand into new areas, either through
research and development or by purchasing startups? Infinity on the
Internet is expressed in terms of the range of goods for sale and the endless deals, offers, and contests. The online consumer is pressured to keep
up with the latest witticism on Twitter, the most entertaining gossip on
Facebook, and whatever is trending on Google. As Dorsen concluded, the
Internet is "clogged with unsignifying noise" and "our political commons

[is] filling up with strings of repetitive trash. We suffer from nonstop agitation and fatigue."[63] In *The Age of Surveillance Capitalism*, Shoshana Zuboff warned that system-wide "information corruption" is built into the business models of companies like Google and Facebook. They have an economic incentive to spread any content that increases traffic on their sites, including sensational claims in advertising and disinformation posing as news.[64] They collect comprehensive data on behavior to captivate and nudge consumers. The leading corporations of surveillance capitalism seek to shape perceptions, control consumption, and profit on social relations. Within this system whatever proves exciting becomes part of the competition for consumer attention.

Meanwhile, during the quarter century when the Internet spread and forged links with the physical environment, VR continued to improve its simulations and illusions of immersion, in which a user temporarily accepts its version of the world. As VR became part of military gaming, activists also began to use it. For example, in *Darfur is Dying*,[65] the player pretends to be a refugee who must navigate around death squads to get firewood, water, and other necessities. Such VR seems far more real than watching television, which by comparison is "a low resolution, sampled, duplicate of reality that does not cater to all senses."[66] When people are immersed in VR, they may take ownership of a virtual body, learn to manipulate an additional limb, or feel at home in an alien environment.[67] VR might be used to understand the world better and to sharpen the senses. Jaron Lanier notes that after immersion in VR one has a heightened appreciation for the subtleties of the actual world. Like a traveler who has lived for a time in another country, the visitor to VR returns to a real world that seems refreshed and defamiliarized. In Lanier's words, "Everyone becomes used to the most basic experiences of life and our world, and we take them for granted. Once your nervous system adapts to a virtual world, however, and then you come back, you have a chance to experience being born again in microcosm."[68] These effects may be caused by what is missing in VR, which is largely a visual and aural construction, supplemented by the sense of touch in only some parts of the body. Some of the senses are disengaged. During a VR visit to a recreated Grand Canyon one cannot feel the heat of the sun on one's face, enjoy a cooling wind, or smell a nearby juniper tree. Nor does VR by itself provide the engagement with the land that can only come from living there and knowing its stories. After extensive experience with the Navajo and study of their oral tradition, Terry Tempest Williams concluded that "each landform, each significant site, seemed to have a name accompanied by a story. These stories animated the country, made the landscape palpable and the people accountable to the health of the land, its creatures, and each other."[69]

VR cannot provide that sense of place, but it can expand our sense of the world. Lanier argues that through VR human beings can increase the depth of their communication. Experiments show that one can learn to control extra arms and legs in virtual reality until they seem a familiar extension of one's own body. Human beings also can learn to control an animal avatar's tail or to "inhabit" a lobster, demonstrating greater psychological plasticity than one might have imagined. A biologist at Cal Tech, Jim Bower, suggested that in some cases the brain might be "recalling" how to control body parts from the phylogenetic tree of the apes that evolved into human beings. Lanier speculated that avatars might even "foretell creatures the brain might be pre-evolved to inhabit in the deep future."[70] The brain seems able to become comfortable in quite different bodies, and it can temporarily morph to different forms, colors, and textures. One VR program, *In the Eyes of the Animal*, takes users to Britain's Lake District where they experience "the perspectives of different creatures, from midges to frogs to owls," simulating their perceptions. Rather than tell a story, it opens a field of experiences and invites the user to explore. The midge provides a microscopic view; the frog swims underwater and explores the margins of wetlands; the owl flies over the same landscape. Instead of the rapid cuts of filmmaking, such VR technology can "slow us down and make us wonder again."[71] Lanier hopes that VR will make possible avatars that can transform their appearance as radically as an octopus can. With malleable bodies, one could express far more than a human body can, enabling a "post-symbolic" form of communication where the mediation of words would often not be necessary. "Suppose we had the ability to morph at will, as fast as we can think. What sort of language would that make possible?" Lanier does not foresee the disappearance of language, but the development of new forms of expression that are not abstract. This kind of VR explores new forms of experience, for example swimming like a dolphin or flying like a swallow, including having virtual eyes that see what a dolphin or swallow can see.

Such projects might require new forms of computer programming. Most programming, as Lanier points out, has rigidities locked in, so that images and sounds lack nuance. Most software was not created for the purpose of rendering subtle differences, and the architecture of this software is built into many programs as a standard element. For example, MIDI is the software commonly used to represent and to play back musical notes. Because it is based on the piano keyboard, it cannot reproduce the "bent" notes on a blues guitar or the subtle sound variations possible on a violin. As Lanier has noted, "Before MIDI, a musical note was a bottomless idea that transcended absolute definition."[72] It varied from culture to culture, and even between orchestras. One orchestra might choose to tune to A440, another to A435. In contrast, MIDI rigidly defines musical notes, and it is entrenched in computers and phones. Given its limitations, MIDI should not be part of any VR project. Subtleties of color are also hard to reproduce digitally. Art historians disdain reproductions of paintings on computer screens because they cannot capture the nuances created by layers of oil paint. It would also be difficult to provide a computer that matches a chef's sensitivity to taste. Much of what one sees and hears through digital systems has been robbed of complexity to compress the size of data files. Lanier calls this "digital reification," which eliminates the fluidity and indeterminacy of actual experience.[73] MIDI and similar programs are roadblocks to creating VR that is close to human experience, including the experience of obscurity that is important in the sublime. Moreover, to create VR that matches the sensory capacities of a human being it is not enough to replicate each sense separately. People routinely synthesize colors, fragrances, tastes, sounds, and textures into a pattern. In the sublime, this synthesis builds to an overwhelming experience.

The challenges to creating fully convincing VR remain massive. For example, a frequent visitor to Niagara Falls discovered that even a slight change in position, from leaning over a railing to standing up straight, radically changed its soundscape. Standing one hears "the roar of the waterfall like something nearby; it's over there, I'm here. But if I put my ears forward of the railing and cliff face, the roar encompasses me. I'm inhabiting a dense soundscape, hollow and continuous and full of subtle variations."[74] Perhaps one day VR will be able to replicate such variation for all nine senses, but even if it does, the real Niagara or Grand Canyon will not be the same from one day to the next.

Aside from the technical limitations that VR designers are working to overcome, can virtual reality escape being captured by Amazon, Apple, Facebook, Microsoft, and Google? These corporations are investing in VR, acquiring startups, filing patents, and courting content developers.[75] VR is more than an upgrade of the eighteenth-century panorama, but it might merely provide titillating stories to draw viewers to additional purchases, based on whatever excites eye movement or accelerates heart-beats and breathing patterns. Kant imagined such emotional responses as the beginning of a transcendental experience, in which Reason played the central role. But in commercialized VR, the subject has emotional experiences that are immediately explained by the software program. Often, VR is being overlaid on an existing room or landscape, to create a mixed reality, or MR.[76] Will MR enhance or vitiate sublime experiences?

Burke noted that in the natural sublime the mind is "so entirely filled with its object, that it cannot entertain any other, nor by consequence rea-son on that object."[77] Confronting the natural sublime, Kant's subject felt insignificant before the scale and manifest power of the universe. Experi-encing VR could evoke the intensity of experience that Burke notes, and it might have mental consequences akin to those Kant described. But just as not every tall building is sublime, most VR is not sublime. It has rap-idly been adapted for use in architecture, real estate, tourism, shopping, psychology, medicine, law enforcement, and conference presentations. But some VR is more ambitious, notably the Cave Automatic Virtual Environment, or CAVE system. For example, at Los Alamos, visitors can enter a CAVE chamber and experience a simulated atomic blast, which cannot otherwise be witnessed because above-ground testing was banned in 1962. In this CAVE "one stands 'inside' a nuclear explosion wearing 3D glasses."[78] No one would survive the experience in the physical world, and the heat, shock waves, lighting effects, and winds are puny compared to an actual blast. Few scientists or politicians born after 1950 have seen an actual test or even the giant craters that remain at the Nevada Test

Site. Because of testing restrictions, new designs of atomic weapons can only be studied through simulations.[79] Unlike a rover on Mars, which demands an interactive relationship with the team driving it, the bomb simulation is a recording, a repeatable experience. CAVE sites are becoming numerous, and they will increasingly be able to simulate experiences of the sublime.

VR haptics also is advancing rapidly. In November 2019, a research team from China and the United States presented an epidermal VR system based on soft sheets of material containing mechanical vibratory actuators that wrap on to the body.[80] The sheets sandwich between layers of silicon actuators, radio frequency loop antennas, resistors, capacitors, rectifiers, and switches. The sheets can simply be wrapped onto the skin and later peeled off. They provide a "skin-coupled haptic interface" unencumbered by external wires or cords, enabling both free movement and high-quality simulations transmitted through the skin, "a relatively unexplored sensory interface." Such new technologies will improve the quality of VR and make it easier to use. It might also improve the tactile experience of scientists driving rovers on other planets. These applications will not necessarily be sublime. However, they might allow people who now can only see and hear one another through a computer to simulate touching one another as well, suggesting a wide range of new applications.

Nevertheless, as of 2021, few digital experiences are sublime. Surfing the Internet is not. Games provide more coherent and absorbing experiences, but they regulate player behavior and overdetermine experience. Many have been developed for military applications, but the sublime is not about winning, losing, or killing. VR is more promising. Yet digital representations also have disqualifying flaws. Most obviously, the colors, shapes, and sounds are compressed and less nuanced than those of direct experience, though advances in haptics can improve the tactile dimension, which increases the power of the experience.[81] Even without enhanced haptics, those enjoying VR often feel that the boundary between self and machine has dissolved, in a suspension of disbelief that, because of interactivity, is stronger than when viewing a film on a large screen.[82] Yet, the powerful stimulus of real danger is absent. Furthermore, some VR users suffer headaches, nausea, retching, and disorientation, and these symptoms can become more acute during prolonged exposure.[83] A similar problem emerged in the 1980s among pilots being trained using flight simulators. Note too, that VR is not an open-ended experience but will eventually reach limits where the simulated world ends. Experiencing an actual place, such as a mountain or waterfall, is endless and unpredictable. It varies greatly depending on who is present, the time of day, and the weather. Each time is unique, as anyone discovers who makes return visits to a natural site such as the Grand Canyon, a Norwegian fjord, or Mount Fuji. By comparison, in VR, places remain the same. Barry Lopez has argued, "One can never, even by paying the strictest attention at multiple levels, entirely comprehend a single place, no matter how many times one might travel there. This is not only because the place itself is constantly changing but because the deep nature of every place is not transparency. It's obscurity."[84]

Lopez understood what Burke also noted: the sublime arises in part out of obscurity. A precise drawing of a palace or a magnificent landscape may provide "a very clear idea of those objects" but does not stir the viewer. Burke considered a "lively and spirited verbal description" better than "the best painting" when seeking to arouse the passions, precisely because obscurity strains the imagination toward infinity. "In reality, a great clearness helps but little towards affecting the passions, as it is in some sort an enemy to all enthusiasms whatsoever."[85] He concluded that obscurity is a powerful stimulus of the sublime. VR seeks to attain clarity and precision, however, and it will need further adjustments to become enticingly obscure.

Like the intangible sublime, VR is based on technological mediation, and it does not involve immediate danger. But there are vital differences. The intangible sublime aims at knowing the truth about black holes, the behavior of subatomic particles, and other scientific questions. These open-ended investigations of the physical world, like the Mars rover, may lead to unexpected discoveries. Civilian gaming, Internet sites, and most VR are dedicated to entertainment. They use a limited sensorium to offer standardized illusions. Some VR does more than immerse one in an environment, however, allowing users to become embodied as nonhuman avatars with unique capabilities. Such VR offers fundamentally new experiences. It may in some cases be sublime, even though it cannot offer the obscurity, complexity, and unpredictable daily variations of physical landscapes.

Yet experiences of the digital sublime are still in their infancy. They need not be commercialized or be tied to the Internet, and they are evolving into new forms of art and interactive experience. Their sublime potentialities are still emerging. Recent research in Canada and Taiwan discovered psychological benefits for people who experience VR simulations of nature. Canadians who wandered in a virtual forest relaxed and their heart rate fell. Taiwanese who visited a virtual national forest had reduced "confusion, fatigue, anger-hostility, tension, and depression." While VR cannot deliver a forest's textures, smells, tastes, or natural phytoncides that boost the immune system, it has clear benefits. Even game landscapes occasionally recreate natural scenes, including "towering redwoods, rolling valleys, and vast, star-filled skies," and these sights can evoke a sense of wonder. A postdoctoral researcher in Milan, Alice Chirico, found that people who are awed and overwhelmed by such vast scenes improve their mental health. She concluded, "This diminishment of the self isn't just a way to feel annihilated, it's a way to find your place in the universe."[86]

VR can potentially stimulate curiosity about companion species and the environmental sublime.

7

ENVIRONMENTAL

While some scientists use digital technologies to access otherwise intangible realms such as distant galaxies, others study the symbiosis of ecological systems and the threats posed by pollution, resource exploitation, global warming, and overpopulation. The exploration of other planets appeals to astronomers, but what are the prospects for planet Earth? Bill McKibben notes that our genetic inheritance links us directly "with every human that came before." We can "look at rock art carved into African cliffs and French caves thirty thousand years ago and feel an electric, immediate kinship." Our sensorium is the same as theirs: "we still hear in the same range and see in the same spectrum, still produce adrenaline and dopamine in the same ways, still think in many of the same patterns."[1] The fascination with such art also includes curiosity about what the animals and plants it depicts meant to ancient humanity. Our sensorium may be the same, but our understanding of nature has changed.

Those studying ecology see human cultures as part of the natural world and cultivate a form of awareness that can be called the environmental sublime. Christopher Hitt noted two decades ago that the classic sublime of Burke and Kant contained elements that often had been overlooked. Both emphasized that the sublime forced one into humility, in recognition of human insignificance before the irresistible forces of nature.[2] The problem with the classic sublime is not the initial sense of weakness, humility, and wonder, rather it is with the subsequent assertion of Reason, when humility "is transformed into self-apotheosis, validating the individual's dominion over the nonhuman world." The environmental sublime rejects this aspect of the natural sublime and instead sees the sublime moment as a recognition that nature is a living, recalcitrant realm: "a sublime encounter with nature seems to have the power to jolt us momentarily out of a perspective constructed by reason and language."[3] Unlike the technological sublime, this sublime is not about the triumph of human reason over natural obstacles and forces but about the ineffable complexity of nature, understood not as a catalog of objects but as an often obscure skein of symbiotic relations.

The environmental sublime retains the humility of the natural sublime but replaces the Kantian apotheosis of Reason with patient immersion. Like the previous sublimes discussed, this one also has two modes. The immersive mode emphasizes symbiosis, and it arises as an awed realization of the almost infinitely complex relations between living organisms. Even a single ant hill contains millions of bits of genetic information. Understanding the ants in relation to their environment is an enormous task, and yet it is only one small part of understanding an ecology and the relationships embedded in it. The traditional natural sublime with its apotheosis of Reason could be enjoyed from the rim of the Grand Canyon or Niagara Falls. In contrast, the symbiotic view of the sublime understands the landscape neither as a striking view nor as a wilderness separate from culture, but rather as overlapping systems of relationships between geology, climate, plant and animal species, and human beings. The view may or may not be uplifting, but the ecological visitor wants to know how much air pollution blurs the details, what species are endangered or already extinct, and how the ecosystem has changed over generations. In the case of the Grand Canyon, during the last century several species of fish have disappeared, wild burros and buffalo have been introduced, and many native plants compete with nonnative invaders. The water temperature and nutrients in the Colorado River have fallen since hydroelectric dams blocked its flow. The dams trap sediment, and the river's color has changed from muddy brown to blue. Its regulated flow has weaker erosive effects.[4] In short, the environmental sublime may still begin with a jolting encounter, but it only emerges fully from a detailed and nuanced understanding of a site. It focuses just as easily on preserved remnants of the tall grass prairie as on the spectacular landscapes such as Yellowstone or the Grand Canyon. It relies on both personal experiences and on scientific instruments to determine the acidity of the water, air quality, remnants of pollution in the soil, and other characteristics of the site. In this understanding, sublimity resides less in a grand view than in the complexity of life enfolded within it, including human cultures as part of the pattern.

At times, nature reappropriates human constructions. Andrew Moore's photographs of Detroit after many of its factories closed and the population shrank "represent a transformation in the use of nature in the Rust Belt from the raw material of industrial manufacturing to the aesthetic of a postindustrial sublime." His images "imagine an environmental renewal by invoking an unexpected sense of nature, conveyed through sharp contrasts between rigid ruins and protruding plants. The result is a sublime vision that gives nature the authority to transform the image of Detroit into a novel, yet disturbing landscape that blurs the lines between wilderness and the city."[5] More often, human actions threaten ecologies, and lead to the shocked realization that a species of insect, an animal, or even an entire ecological system, is blighted or in danger of extinction. A new form of fear emerges, for the elimination of even one species can have severe consequences. In one location, the extinction of the sea otter meant that "sea urchins proliferated, reducing the amount of kelp and seaweed that is the sea urchins' chief food" until they largely died off.[6] The results are similar whenever a "keystone species" is eliminated. Deer, freed from wolves, proliferate, overgraze, and then starve to death.

The second form of the environmental sublime emerges out of this concern with symbiotic complexity. Where earlier formations of the sublime incite a feeling of mortal vulnerability when facing a volcano, a disaster, or a battle, the environmental sublime arouses a fear for the fate of a species, habitat, or ecology. This form of the environmental sublime arises with the realization that "the wounds of the natural world are also social wounds, and that the planetary ecological crisis is the material and historical consequence of an anthropocentric and dualistic worldview."[7]

Unlike earthquakes or other abrupt disasters, habitat destruction and species extinction may unfold gradually, as a form of what Rob Nixon has called "slow violence." The world's coral reefs have been dying for decades; the flightless Great Awk gradually disappeared over centuries before its extinction. Similarly, Jeffrey Bolster's *The Mortal Sea* examines the protracted decline of the fishing industry in the Atlantic Ocean during the thousand-year age of sail. After Europeans depleted the eastern side of the ocean, they were astonished by the abundance off the coasts of North America. As early as 1639 colonists attempted to regulate the fisheries but with little success. Already in 1703, some lamented overfishing off Newfoundland, and in the following generations species after species of fish was decimated, including sea bass, salmon, and shad. George's Bank once seemed to be a limitless supply of halibut, but they were fished out by 1880. In New England's fishing grounds there was a slow collapse of cod, mackerel, menhaden, halibut, and lobster stocks. In each case, an ever-smaller supply was caught ever more efficiently.[8] Bolster demonstrates that throughout this long process some fishermen and officials understood the dangers but nevertheless proved unable to halt the destruction of one fishery after another.

The environmental sublime concerns such failures to maintain ecologies. It also concerns the loss of knowledge that occurs when aboriginal cultures are overrun and transformed. Barry Lopez notes that by the early nineteenth century many natives of the Arctic had been decimated by disease. British whalers on their annual return found whole villages empty. Fortunately, the Inuit were not entirely wiped out. They revived. But today, they are threatened by oil and mining interests and by new shipping lanes that global warming has made possible. Lopez argues, "the element of the eco system at greatest risk is not the bowhead [whale] but the coherent vision of an indigenous people." They alone have a long and intimate knowledge of the Arctic, plus the resourcefulness and economy of action that enables them to survive in that difficult environment. "Our intimacy [with that region] lacks historical depth and is still largely innocent of what is obscure and subtle there."[9] We remain outsiders, stunned by the Arctic as a vast, natural sublime. We effect great changes, unaware of the region's complexities. Lopez notes that the Inuit, "who sometimes see themselves as still not quite separate from the animal world, regard us as a kind of people whose separation may have become too complete," becoming "'the people who change nature.'"[10] To the Inuit, the animals, plants, and land are not objects. They "have difficulty imagining them-selves entirely cut off from the world of animals. For many of them, to make this separation is analogous to cutting oneself off from light or water."[11] Something similar can be said for most aboriginal peoples.

The European sublime view of nature became prominent after around 1750, during the industrial separation from nature. At first the sublime was not split into many formations. In the early nineteenth century, both poets and geologists used the same rhetoric when describing their sublime experiences of nature. In 1805, when the young Humphry Davy lectured in London on volcanoes, he cited first-hand accounts of Italian volcanic eruptions. He said, "language must necessarily fail when applied to such a purpose. And not even the most perfect delineation of the most perfect artist could do justice to the combination of circumstances in which feel-ing, and hearing, and sight are almost equally concerned, in which the earth trembles, in which the continued sound of thunder dwells upon the ear, and in which the eye is constantly dazzled by lightning flashing above and by liquid fire streaming below." He considered a volcano "the most sublime" of geologic phenomena.[12]

English romantic poets hoped the experience of the sublime would reunite the human mind and nature, but this unity of perception was eroded during the following two centuries. Many critics have pointed out[13] that Wordsworth's encounter with Mont Blanc in *The Prelude* did seem to describe such a unity, and his encounter with its overwhelm-ing power was followed by a successful effort to use his imagination to rise above and make sense of the experience. This resembled what Kant described as the awakening of Reason. To that age it meant that nature was not "other," and that human beings were not "other" to nature, either. Rather, it seemed that encounters with nature could awaken inner powers. As Ralph Waldo Emerson once put it, "mountains are silent poets, and a view from a cliff over a wide country reinstates us wronged men in our rights. The imagination is touched."[14] Looking back from the present, however, the Kantian sublime and the sublime of nineteenth-century Romantic poets were not only about nature but also about the power of the mind to subsume it. Visitors to Niagara Falls or the Grand Canyon felt overwhelmed by the thundering cataracts and enormous vis-tas, but these experiences were seldom based on much knowledge of the botany, geology, or history of the site. In the encounter, nature was not "known."

The understanding of nature underwent a revolution during the nineteenth century. Wordsworth knew the Lake District where he resided for much of his life; and Emerson likewise knew eastern Massachusetts well. Both studied nature to understand humanity. But their romantic vision began to seem quaint as geology developed a new language to explain the landscape and pushed back the history of Earth millions and then billions of years. Charles Darwin further revealed this new relationship to nature. Human beings turned out to be recent evolutionary arrivals who appeared only during the last one percent of Earth's history. After these humbling discoveries, the sublime seemed less a revelation of unity with nature and more like a cultural construction.

The perception of a widening divide between human beings and nature helps to explain the appeal of the technological sublime. During the nineteenth and twentieth centuries, many celebrated the domination of nature, as demonstrated by canals and railroads that conquered space, the acceleration of communication that conquered time, the great dams and skyscrapers that triumphed over gravity (further dramatized by balloons and airplanes), and the overthrow of darkness through electrification. This domination became a source of pride, and some nations began to identify themselves with their tallest buildings, largest dams, and most impressive bridges. At the same time, human beings increasingly acquired knowledge through mediating technologies, in roughly the following sequence: the telescope, microscope, panorama, photography, scale models at world's fairs, amusement park reconstructions, films, radio, television, computers, satellites, and extraterrestrial rovers. These experiences were at one remove from direct physical contact.

During this process, the experience of the sublime was stripped of its metaphysical aspect. In Wordsworth or Emerson the sublime had a transcendental meaning, and, as Brady notes, because of its religious and pantheist associations it became unacceptable to many later Anglo-American philosophers.[15] Yet, as she points out, "Although some eighteenth- and nineteenth-century ideas associate sublimity with God's power as symbolized in nature, in various theories, such as Kant's, we find a more secular sublime, and one which can be called upon for developing a more

pluralistic conception of the aesthetic valuation of nature."[16] As understood in the work of Ronald Hepburn, metaphysics is inherent in the
experience of the sublime, rather than being appended after the fact.
Metaphysics is "an element of the concrete present landscape: it is fused
with the sensory components, not a meditation aroused by these."[17] One
becomes aware of a complexity that is beyond our grasp, or as Brady
expresses it, the "metaphysical imagination does not align with nature
as somehow fully *known*."[18] Brady therefore argues that Kant's sublime
remains viable. This sense of ungraspable complexity is a precondition
for developing a respect for other species and environmental awareness.

Yet that older sublime, while necessary, is not sufficient. The environmental sublime provides an additional way to think about nature in
terms of symbiosis and the dangers of extinction. It does not replace the
natural sublime, which persists in common experience. Brady is surely
correct to argue that it provides "a distinctive kind of aesthetic judgement
and value grounded in tremendous qualities, complex emotions, and an
active, expanded imagination."[19] However, botanists and biologists doing
fieldwork have another kind of experience that emerges from intensive
study of a single place, often focused on one species. *The Ants* by Bert Holldobler and E. O. Wilson is an excellent example, showing in hundreds of
detailed pages how ant colonies function. Holldobler and Wilson identified the "chemical vocabulary" that ants use to communicate. Genetic
research has likewise added to our understanding of plants, animals, and
the ecological systems they inhabit. Like physicists and astronomers,
biologists use specialized technologies to understand intangibilities, such
as the traces of a toxin in water or radioactivity in soil that cannot be
seen, smelled, or tasted.

Environmental scientists work in laboratories, but they also are
immersed in the details of particular places where through patient observation they discover symbiotic relations. In the western Pacific Ocean,
"coral reef damselfish tend underwater algal gardens, where they remove
less desirable algae species and chase away predators."[20] Japanese marine
biologists discovered a species of hermit crab that inhabits a living piece
of coral, which it transports across the ocean floor, making it a "walking
coral." The coral avoids being covered by sand and gets transported to
good locations, while the hermit crab gets protection (figure 7.1).[21] In

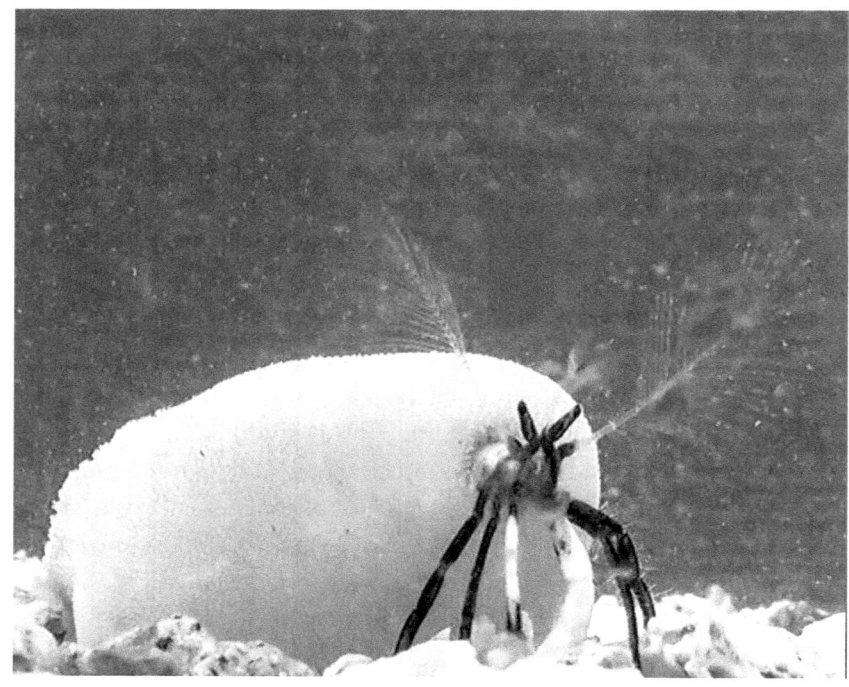

7.1 This coral recently discovered on the coast of southern Japan protects a hermit crab, which lives inside it. In exchange, the crab prevents the coral from being covered by sand and transports it to favorable locations. Courtesy of The Conversation and Creative Commons, https://theconversation.com/newly-discovered-hermit-crab-species -lives-in-walking-corals-84389.

Kenya, scientists discovered that underground termites "increase grassland productivity and biodiversity over large areas by raising soil fertility."[22] Scientific immersion in specific habitats documents how symbiosis permeates the natural world.[23] Comprehending such relationships can be a sublime experience. Just as there is a large reading public for books that explain the origins of the universe and for photography of previously unknown phenomena in outer space, there is a strong interest in accounts of fieldwork. Henry David Thoreau and John Burroughs were forerunners of this genre that includes Aldo Leopold, whose emphasis on biodiversity and ecology are characteristic of the environmental sublime. More recent examples would include Jane Goodall whose lifelong study of chimpanzees showed that human beings are not the only animals that have personalities, can reason, use tools, be altruistic, and experience joy and sorrow. Her experiences of interspecies communication were both scientific and emotional, exemplifying the immersive fascination of the environmental sublime.[24] Subsequent research found that chimpanzees outperformed human beings in some memory tasks and that groups of chimpanzees manifested cultural differences.[25] Goodall demonstrated that chimpanzees in their native habitat behaved far differently than those confined in zoos.

Other fieldwork has yielded evidence that two species of dolphin have cultures that use tools, form abstract concepts, and communicate vocally.[26] Their behavior does not appear to be instinctual or genetically determined. Rather, it seems to be learned, cultural behavior. Nor is the environmental sublime restricted to studies of insects, shellfish, and animals. In *The Hidden Life of Trees: What They Feel, How They Communicate*, Peter Wohlleben combined his experience as a forester with recent scientific research to show how the tangled root structures beneath the ground enable an information exchange among trees, which should be regarded not as isolated individuals but as a biotic community.[27] This community includes more than trees. Botanist Robin Wall Kimmerer writes in *Braiding Sweetgrass* of how "mycorrhizal symbiosis enables the fungi to forage for mineral nutrients in the soil and deliver them to the tree in exchange for carbohydrates."[28] Kimmerer, who is both a university scientist and a member of the Potawatomi Nation, draws parallels between human relationships and the relations developed within groves of trees. Potawatomi stories correctly suggested that trees communicate with one another.[29] Indeed, in the case of quaking aspens, what appear to be many individuals may be a single organism. Suzuki and McConnell note: "In Utah, a single aspen plant made up of 47,000 tree trunks was discovered."[30] Through its extensive root system, it shared water, minerals, and nutrients. All of these studies show the value of patient immersion in ecologies rather than relying entirely on observing animals in a zoo or working with samples in a laboratory

However, the immersive form of the environmental sublime is only half the story. Thousands of species have been marginalized or wiped out as agricultural uniformity spread across the landscape. As Diane Ackerman put it, a "monotony of genes rules much of the planet, as wild and varied habitats give way to more prosperous if homogeneous big farms."[31] At the same time, global warming is decimating some species and forcing others to migrate toward more temperate regions. As Louise Economides explains in *The Ecology of Wonder in Romantic and Postmodern Literature*, "environmental sublimity differs profoundly from its romantic sources insofar as human-made phenomena—such as human-induced environmental catastrophes—form a new locus of fear and awe, rather than natural phenomena."[32] When studies reveal that the northern half of the Australian Great Coral Reef is bleaching and dying or that the population of chimpanzees is rapidly declining, observers may experience an even stronger sorrow than seeing fire destroy a city. A city can be rebuilt, but no extinct species has yet been recovered. When the actions of human beings cause massive destruction, the environmental sublime contains elements of the disastrous sublime and may become a repudiation of the technological sublime.

When the destruction or the extinction is local, however, there often is support for recovery. For example, by the twentieth century salmon had disappeared from most of New England's rivers. The federal government supported programs to reintroduce them, with partial success. Annually, about a thousand wild salmon return to spawn, mostly to Maine.[33] Such restoration projects are more difficult than they look. A study of 621 of them found that only 25 percent achieved their goals.[34] Restoration work expresses a longing for recovery of lost landscapes and ecological relations. Such losses can provoke emotions of grief, guilt, anger, and fear, but they also can strengthen the determination to restore a habitat, whether wetlands, hedgerows that shelter pollinating insects and small animals, or forests. Dolly Jørgensen has examined efforts at rewilding in Norway, which twice imported muskoxen from Greenland and released them. The first shipment was killed by poachers and possibly also by German soldiers during World War II; the second has expanded to about two hundred individuals.[35] After years of controversy, muskoxen have been accepted as a part of Norway's recovery of parts of its preindustrial nature. Likewise, wolves were reintroduced into Yellowstone National Park and accepted after some controversy. In contrast, Denmark reintroduced wild swine, but then, fearing swine flu and other illnesses that might decimate its valuable pork industry, the government exterminated them and built a fence along its German border to keep them out.[36] The technological sublime celebrated the conquest of natural objects and forces. From the perspective of the environmental sublime, however, that conquest includes a sixth wave of species extinction and calls for the protection or recovery of preindustrial nature.[37]

Advances in DNA research could make possible the "de-extinction" of animals last alive in the Pleistocene, brought back because they "belong" to a lost landscape that they might help recreate. This may seem an attractive idea, but it assumes absolute human control of the rest of nature. "Thus, the sublime of catastrophe is succeeded by the giddiness of omnipotence."[38] The concept of de-extinction suggests that human beings can create life forms or turn back the clock, for example recreating the woolly mammoth that disappeared four thousand years ago, as if only one temporal system and one landscape were involved. In 2021, a Harvard research team received $15 million to try to recreate a mammoth in an artificial womb. If successful, it will be a creature with no parents to assist it, based on a combination of recovered mammoth DNA and elephant DNA. The aim is to repopulate Siberia with these creatures who might help convert vast thawing areas of moss into grassland. Another de-extinction project would resurrect the passenger pigeon in the United States.[39] Until the 1870s billions of these birds literally darkened the skies when they migrated between the Rocky Mountains and Ohio. But most of the forests and grasslands that supported the vast flocks are gone. Bringing back an individual passenger pigeon or a mammoth might be possible, but can their flocks, herds, and ecological systems be recovered? This is not a question imagined in the natural sublime of Kant or Burke.

As Nixon has emphasized, every species, including human beings, can be said to "inhabit multiple temporal orders that often coexist in frictional states, shifting and sliding like tectonic plates." There are the times of the seasons, of the tides, of migration, of electoral politics, of interest rates, of factory production, of hurricanes, of global warming, and of the decomposition of toxic chemicals, to name a few. These temporal flows constantly interact. Is a drought the result of global warming or part of a long-term weather pattern? Are bees dying out because of pesticide use or some other cause? Such questions demand, as Nixon phrases it, that we learn to see "the lineaments of slow terror behind the façade of sudden spectacle."[40] The environmental sublime concerns both spectacles confined to a restricted temporal frame and complex, slower temporal patterns and spatial relations.

Scholars argue about whether the Anthropocene began in antiquity, during the industrial revolution, or in the twentieth century.[41] The geological evidence suggests it began before the twentieth century, but the Anthropocene is also a matter of awareness. Human beings long had only a limited grasp of the environmental effects of agriculture or deforestation, and the dangers of industrialization were at first poorly understood. Awareness of these destructive effects can be traced back to the eighteenth century,[42] and it was powerfully articulated in works such as George Perkins Marsh's *Man and Nature* (1864) and Eugène Huzar's *La Fin du monde par la science* (1855).[43] Both wrote of species extinction and the destruction of landscape that had already occurred before 1850. Recent research further documents such destruction. We now know, for example, that before the Spanish arrived the Amazon basin was not a wilderness but a heavily populated region that had developed effective recycling and farming techniques.[44] In the sixteenth century European diseases decimated these communities.

During the industrial revolution, the technological sublime became prominent, but after c. 1965 the realization gradually took hold that human interference often threatened not just one or another species but entire civilizations and ecological systems. In the reforested Amazon basin, by 2018 one of the largest iron mines in the world had produced 1.4 billion tons of high-grade ore,[45] leaving scars that could be seen in Landsat satellite photographs. Aerial photographs taken of Paraguay reveal 55,000 square miles of deforestation in the Gran Chaco plain, visible as hundreds of rectangular areas of clear-cutting (figure 7.2). To survive in the Anthropocene, human beings will need to move beyond celebrating such mines and land clearance, and instead develop a creative engagement with the environment. We need a new way of being in the world. Roy Scranton describes this as "learning to die in the Anthropocene:" In the next millennium, the world will become "unrecognizably different from the one we have known for the last 200,000 years. For us to adapt to this strange new world, we're going to need more than scientific reports and military policy. We're going to need new ideas. We're going to need

7.2 Deforestation in Paraguay, 2016, revealed by the Landsat 8 satellite. The Gran Chaco plain is covered by a dry forest of thorny trees, shrubs, and grasses. Since 1985, 55,000 square miles have been cleared for farming. Courtesy of NASA, https://earthob servatory.nasa.gov/images/92078/deforestation-in-paraguay.

new myths and new stories, a new conceptual understanding of reality."[46] We will need to rethink the sublime. Otherwise, it will remain part of the misleading dualisms of "humanity and nature" or "nature and culture."

During the Cold War these dualisms began to break down both in academia and among the public. As Finis Dunaway notes, during the 1960s environmentalists reacted against government experts who "sought to discredit fear and anxiety as illegitimate emotions" when it came to "technology, the environment and human health. They denied the vulnerability of permeable ecological bodies and urged the public not to be afraid of bomb testing" and to accept the growing use of pesticides. Resistance to these messages could be found not only in print (notably Rachel Carson's works) but also in the imagery of the environmental movement. "Visual images played an active role in this struggle and ultimately helped to legitimate a new emotional style in American public culture." The public gradually learned to see "through an ecological lens" as part of a "shift from the idea of nature as a realm separate from human society to the notion of environment as an interconnected system that all things— human and nonhuman—shared."[47] Often images rather than arguments made this case forcefully, including an astronaut's iconic image of Earth rising over the moon or a 1979 photograph of the enormous cooling towers at the Three Mile Island nuclear plant during its partial meltdown, with a children's swing in the foreground. These images had an emotional impact, as did Al Gore's film, *An Inconvenient Truth*. Such "environmental icons create a picture of universal vulnerability to enframe all people as equally susceptible to ecological harm."[48]

Awareness of environmental vulnerability has also transformed university curricula, notably in the rise of the environmental humanities.[49] Many academic disciplines now reject the once common division between culture and nature. The German sociologist Ulrich Beck concluded, "Anyone who continues to speak of nature as non-society is speaking in terms from a different century, which no longer capture our reality. In nature, we are concerned today with a highly synthetic product everywhere, an artificial 'nature.' Not a hair or a crumb of it is still 'natural,' if 'natural' means nature being left to itself."[50] Likewise, historians have rejected the category of wilderness, as the reification of an ideal state of purity that for a century was simulated by expelling first peoples from their homelands, as occurred with African nomads, Australian Aborigines, and Native Americans.[51] Likewise, geographers and regional planners have abandoned the assumption that the city and the countryside are opposites. Cities have always been ecologies, and it was a modernist fantasy to suppose they were vast machines. Moreover, city residents often have a smaller carbon footprint than suburbanites.[52] In a sustainable future both cities and suburban hinterlands will need to become sustainable.

The breakdown of the nature/culture divide strongly suggests that making a firm distinction between the natural and technological sublimes is no longer tenable. Global warming raises the temperature of the oceans, which intensifies hurricanes. Melting Arctic ice raises the sea level, which intensifies coastal flooding. Changing patterns of rainfall lead to drought and forest fires in some areas and to violent storms elsewhere. Hurricanes, fires, and floods were long considered natural phenomena, but in the Anthropocene technology can both amplify and restrain natural processes. When a city is knowingly built on the fault line between tectonic plates that cause intense earthquakes, as is the case with Santiago, Chile, and San Francisco, California, the ensuing disasters cannot be attributed to natural forces alone. In short, what once seemed the forces of nature are now understood to be hybrid forces that are both natural and technological.

Some architects are beginning to find ways to move beyond the nature/culture dichotomy, as human beings seek to intervene and halt ecological destruction. At Syracuse University, the school of architecture asked design studio students to study specific natural sites, create models of their ecology, and plan "technologically sublime interventions." One group of students examined Africa's Lake Chad, once a large body of fresh water covering 25,000 square kilometers, which has been shrinking since the 1970s to merely 1,350 square kilometers. Neighboring nations have been extracting water for irrigation. If nothing is done Lake Chad will disappear, and many species will become extinct in that region. The design response, called "The Bloom Living Archive," was to build "blooms" that "generate water from capturing CO_2" from the air, which can then be stored in "pods" and released into the lake. "The entire system can be dismantled and transported to new locations as enough CO_2 is absorbed and water bodies are fully recovered." Such "technologically sublime interventions" work with the landscape, seeking to repair damage caused by previous human interventions.[53] Other projects studied how to create more ice to replace that being lost in the Arctic, or how to deal with sandstorms and desertification in Saudi Arabia. The latter project began with the study of sandstorms and the changing shape of the dunes in a specific area. The students then imagined a way to harness the energies of sand and wind. They designed an ingenious system connected to the local water treatment plant. Some of its water could be infused with a bacterium that "turns sand into sandstone when combined with water, urine, and calcium." This makes it possible to stabilize the invading sand as sandstone formed into structures that trap blowing sand below new rooftop gardens. The forces that were expanding the desert can be used to expand agriculture instead. These projects are symbiotically sublime because they combine the large scale of the mathematical sublime with human ingenuity, to build structures that can restore a damaged landscape. They suggest the kinds of responses needed to move beyond apocalyptic fatalism. Such projects become even more attractive when alternative energy sources drive them.

Both the natural and the technological sublime presupposed that an observer is in a safe location, contemplating a powerful or impressive object. But in the Anthropocene Earth is no longer safe, and human beings can no longer imagine that they are mere observers. The environmental sublime moves beyond contemplation of what is lost or about to be lost, and it recovers a sense of humanity's immersion in the natural world. In contrast to the natural sublime of Kant, this sublime is about "biometric assimilation." The subject realizes a new form of aesthetic pleasure on being "thrust back into a newly realized space that requires tending and care."[54] Or as Timothy Morton underscored in *Being Ecological:* "You are already a symbiotic being entangled with other symbiotic beings."[55] In *Ecology without Nature* Morton criticized Kant because his idea of the sublime focused too much on the interpretation of experience, rather than on the wonder of experience itself. "In Kant's terms, our mind recognizes its power to imagine what is not there: 'Sublime is what even to be able to think proves that the mind has a power surpassing any standard of sense.' Kant demonstrates this by taking us on a journey of quantity, from the size of a tree, through that of a mountain, to the magnitude of the earth, and finally to 'the immense multitude of such Milky Way systems.' The sublime transports the mind from the external world to the internal one."[56] The movement, somewhat ironically, is away from nature and into the mind.

In contrast, the environmental sublime renews intimacy with the tangible world. Seeing the wounds human beings have inflicted, it embraces ambitious restoration projects and works toward ecomimesis. As Lillian C. Woo explains, "Using nature as its template, ecomimesis conserves, repairs, and improves existing ecosystems." The "goal is to design a human community that does not interfere with nature's inherent ability to sustain life in the Earth's biosphere and minimizes disruptions to nature's ecosystems. Its primary goals are to re-establish ecosystem stability, preserve regional biodiversity and habitats through continuity of functions and connectivity, and conserve, repair, and improve existing ecosystems."[57] Woo advocates expanding natural habitats, constructing buildings that have green walls and roof gardens, designing technologies for disassembly and reuse, using low toxic materials, recycling waste, and adopting "living machines" that "produce food or fuels, treat wastes, purify air, regulate climates," and more.[58] Such practices create a sustainable landscape and change the human perception of it. But there is no universal design solution, for the relationships between residents and their landscapes vary. Building a community based on ecomimesis in Siberia will hardly resemble ecomimesis in the Amazon or on the Greek islands. Earth's variety is too extensive for one person to know many places well. Thoreau understood that when he declared that he had traveled widely in Concord. Ecomimesis requires deep knowledge of a home place.

Technology does not offer a way to escape or to conquer nature, because human beings and their cultures are inseparable from nature. The technological sublime was based in part on an implicit belief that consumption had few environmental consequences. It blinded human beings to their effect on other species. The environmental sublime entails a quite different view, in which human beings are not hierarchically above other life forms. Rather, we are a companion species. Since at least the industrial revolution, however, we have been a careless companion, filling the sea with plastic, the air with pollution, and the earth with pesticides and chemical waste. Recognizing the enormous scale of these blights on the world can lead to despair and nostalgic contemplation of what is already lost or about to be lost.[59] The way forward demands that narratives of recovery replace the hubris of the technological sublime, which often has celebrated human domination and destruction of the lifeworld. The environmental sublime instead finds wonder in other species, whether a forest, groups of chimpanzees, or a pod of whales. It does not treat them as raw materials to be "harvested" or as a backdrop to human endeavors, but as a source of knowledge and connection.

The land itself is not passive. As Kimmerer put it, "What knowledge the people have forgotten is remembered by the land."[60] The beans remember how to fix nitrogen and therefore should be planted next to corn and squash. Grown together, these three provide more nutrition than when planted separately. "The gifts of each are more fully expressed when they are nurtured together rather than alone." Maintaining this symbiosis requires a human gardener. "Corn, beans, and squash are fully domesticated; they rely on us to create the conditions under which they can grow. We are part of the reciprocity."[61]

In contrast to other forms of the sublime, the environmental sublime requires an understanding of the multiple temporal rhythms of plants, insects, birds, animals, and the weather as they interact. Seeing landscapes in terms of the environmental sublime demands attention to both the microscopic and the panoramic view. It is not about the conquest of nature, nor is it about ruined landscapes of the disastrous sublime. Rather, it concerns complex relationships slowly unfolding. It contrasts with the frantic acceleration and monumentalism celebrated in the technological sublime, the deadly finality of the martial sublime, and the timeless replays afforded by virtual reality. The environmental sublime locates human beings inside the landscape, where they are not hierarchically superior to other species. This is a sublime that wonders at the complexity of symbiosis and fears humanity's often destructive relation to the rest of nature.

CONCLUSIONS

8

NATIONALISM AND THE SEVEN SUBLIMES

The previous chapters have focused on possible sublime experiences, without discussing how they have been interpreted in different national contexts. Full consideration of this topic would require an additional book, and this chapter can only demonstrate one important point: that sublime experiences are universal, but the meanings nations give to those experiences are not. Each of the seven forms of the sublime is potentially available to anyone. None is exclusive to a particular social class, gender, race, or nationality. The sublime response is universal. Longinus, in his foundational treatise on the sublime, focused on certain literary passages that he averred were universally recognized as being sublime. As Malcolm Heath summarizes, the "genuine sublime gives delight always and to all, and the consensus of people of many different backgrounds provides incontrovertible corroboration."[1] Kant or Burke discussed not only the natural sublime but also disasters, impressive buildings, and the effects of explosions and cannon fire. Based on their writings, the four unmediated sublimes (natural, technological, disastrous, and martial) all have a claim to being universal experiences. Furthermore, it appears that people from every walk of life can have these experiences. These four tangible sublimes therefore have the potential to unite rather than to divide humanity. One does not need to be American to be awed by the Grand Canyon; or French to enjoy the view from the Eiffel Tower; or

Italian to be moved by the ruins of Pompeii; or European to be moved by the battlefield and panorama at Waterloo. Likewise, mediated sublimes appear to be universal. Intangible atomic particles or revelations of outer space are not the exclusive property of one nation. Galaxies, black holes, and the microscopic realms of biology can evoke anyone's wonder. Nor are digital experiences such as virtual reality the exclusive preserve of a particular race, gender, or social class. And as for the environmental sublime, it reveals symbiotic communities and complex dependencies. National borders mean nothing to the Jetstream, the Gulfstream, or the blowing sands of the Sahara. Birds migrate between continents, and satellite photographs depict disappearing forests and dying coral reefs, whose demise affects us all. Mediated sublime experiences disclose not separate nations but international connections. The seven sublimes, taken as a whole, would seem to weaken nationalism and to unite humanity in common experiences and the recognition of a common fate.

In practice, however, particular social classes, nations, and genders have claimed a special affinity for one or another form of the sublime. Even Kant, in his early writings, mistakenly saw the sublime as a male emotion, assigning to women a greater feeling for the beautiful.[2] Certain nations and regions also have invested particular sublimes with nationalist connotations. Indeed, the rise of nationalism in modern Europe occurred simultaneously with the increased interest in the sublime. One foundational text for the study of European nationalism is Ernest Renan's lecture "What Is a Nation," delivered at the Sorbonne in 1882, in which he argued that a merely geographical definition of any nation was insufficient. There is also a national imaginary that combines stories, values, beliefs, patterns of thought, and symbolic sites. Renan's line of thinking was reinvigorated a century later by Eric Hobsbawm and Terence Ranger's *The Invention of Tradition* (1983) and Benedict Anderson's *Imagined Communities: Reflections on the Origins and Spread of Nationalism* (1983).[3]

The sublime is one of many things that can become part of a cultural identity, however. Not all spectacular sights are incorporated into a national imaginary. For example, the Kamchatka Peninsula on Russia's Pacific coast is larger than Italy, and its scenery rivals that of the American West. Its two mountain ranges include 150 volcanoes, more than twenty-five of them active.[4] Its highest peak rises above any in Austria, Germany, or Switzerland. Its hot springs and geysers are as spectacular as those at Yellowstone National Park in the United States, and its 446 glaciers out-number those in famous national parks elsewhere. Its wildlife is diverse and protected, including the largest salmon runs in Asia and the world's largest concentration of brown bears. Yet the Kamchatka Peninsula has not been as important to Russian identity as comparable places were in the United States or Europe. Indeed, during the Soviet era, the military tested missiles there, and tourism was tightly restricted. Only recently have many visited its sublime landscapes.[5]

In the United States, such spectacular scenery became a central ele-ment in national identity. After attaining their independence in 1783, white Americans lacked a long history, a shared religion, or a royal family to invest with nationalism. Instead, they identified the nation with the natural sublime, exemplified by sites such as Niagara Falls, the Natural Bridge of Virginia, Yosemite, and Yellowstone, and with the technologi-cal sublime, exemplified by the Erie Canal, the railroads, and Brooklyn Bridge.[6] In the United States, the natural and technological sublimes became important almost simultaneously, and they developed together. Gordon Sayre concluded that in the decades immediately after indepen-dence, the United States began to embrace the sublime in poetry and in travelers' descriptions of scenery. "The American sublime inspired Romantic idylls of mystery and awe, but also puffed up nationalist pride and invited schemes for mills, canals, and bridges."[7] The technological sublime, in the form of imagined canals and mills, early emerged as a component of nationalism. In the twentieth century, Americans adopted additional sublime sites, including the Grand Canyon and the Empire State Building.[8] As Ulla Haselstein has noted, Americans tended to see the sublime in positive terms, rejecting the idea that the sublime is insepa-rable from "the experience of the inadequacy of the imagination" lead-ing to a "crisis of representation."[9] This negative sublime has been more common in Europe, and it found full expression in the work of Jean-François Lyotard, who saw it as being inseparable from the avant-garde's striving to demonstrate the "incommensurability between thought and the real world."[10]

A different interpretation of the sublime was common in the United States.[11] As the physical world was increasingly demystified by science, the sublime represented a way to reinvest the landscape and the works of men with transcendent significance. Chandos Michael Brown concluded that from Thomas Jefferson to Thomas Cole, or roughly 1780 to 1840, American poets, painters, and travelers constructed a view of the sublime that performed important culture work. This American sublime "yoked together evangelical millennialism and revolutionary republicanism and situated them in a specific environment," and "it emphasized above all the power of environment in shaping not only the somatic self but the collective character of a people." In this view, contact with sublime nature "worked on the plastic stuff of the European mind," creating a new national character.[12] Indeed, Americans concluded that both the natural and the technological sublime inspired and strengthened democracy.

This argument about the formation of a national ideology should not be taken to mean that the United States was exceptional. Americans were not unique in celebrating technological achievements, for example, as other nations saw their constructed environments in similar ways. Yet cross-cultural comparisons strongly suggest that while sublime experiences are universal, the meanings given to them often are not identical. Societies emphasize different sublimes, and even when they focus on similar experiences their interpretations vary. The study of the sublime ultimately cannot be separated from politics and national tensions. Émile Durkheim usefully observed, "The ideal society does not stand outside the real society: it is part of it. Far from being torn between two opposite poles, we cannot be part of one without being part of the other. A society is not simply constituted by a mass of individuals who compose it, by the territory they occupy, by the things they use and the actions they perform, but above all by the idea it has about itself."[13] Powerful experiences of the sublime often become part of national identity. Yet this will seldom be a coherent identity, but rather, as Durkheim argues, a conflicted one. In the United States, both the natural and technological sublimes have long been considered part of the national self-definition, yet not all those who celebrate the Grand Canyon embrace Hoover Dam or the lights of Las Vegas.[14] Not every form of the sublime is woven into every national identity, but most societies embrace both the natural and the technological sublimes to some degree and many of the other forms, especially at the locations of disasters, ruins, and battles. Differentiating between seven forms of the sublime enables a nuanced analysis of national identity.

Consider, for example, the salience of the sublime in Spain, Britain, and Denmark. In none of these nations is the technological sublime as central to national identity as in the United States. The Spanish have a long and rich heritage including Greek, Roman, and Moorish monuments, as well as cathedrals, monasteries, palaces, and sites from the Renaissance when they were the preeminent European power. The national identity as well as conflicting regional identities were well established before the rediscovery of the sublime after 1700. Denmark has an equally long history, with national symbols that include Viking ruins, royal palaces, and cathedrals. They have built few skyscrapers, and their massive bridges are less important symbols than the Rune stones in Jutland or the Little Mermaid statue in Copenhagen's harbor. The British invest a wide range of objects with nationalism, including Stonehenge, Buckingham Palace, York Cathedral, and the ruins of the Roman wall near the Scottish border. Since the early eighteenth century, the British have appreciated the sublime, perhaps more so than the Spanish or the Danes, but it is not the preeminent category of their nationalism. Cian Duffy shows in *The Landscapes of the Sublime, 1700–1830* that British travelers, poets, and philosophers have long thought that the quintessential examples of the sublime were not found in Britain but in the Arctic, the Alps, Italy, and North African deserts.[15] In contrast, Americans celebrated sublime landscapes within their own borders.

British national identity after around 1780 was expressed less through the sublime than through the picturesque, an aesthetic category that blends the sublime and the beautiful. In formal terms, as Jens Jager notes, the picturesque values "irregularity, a mixture of lights and shades, contrast and diversity, such as to be found in ancient architecture, [and] rough but not unpleasant nature." By circa 1800 the picturesque reflected a political preference "in favor of the traditional British system, against the constructed and abstract French" gardens and landscapes.[16] Accordingly, the "English aristocracy acted in expensive ways to match the picturesque ideal (move a village, dig a lake, shift some trees)."[17] As interest in British heritage and monuments increased, it was codified in laws protecting certain sites.

Paintings and photographs often depicted Britishness through images of ruined abbeys, cottages, and rural scenes. J. M.W. Turner (1775–1851) "gave visual form to Burke's categories of the sublime, such as Vastness, Obscurity and Power," in landscapes of both natural sites and "new industrial spectacles, including a train speeding through the countryside in 'Rain, Steam and Speed—The Great Western Railway' (1844)." By circa 1850, "the identification of sublimity with new technological forms, such as railways, bridges and factories, formed a key component in the presentation and reception of industrial development" including drainage systems, viaducts, and other projects depicted in the illustrated press.[18] However, as Paul Fyfe notes, most British "artists adapted the picturesque to represent industrial sites, including collieries, factories, and railways." In their hands, the picturesque "could be made to accommodate industrial forms newly present on the landscape." A railway line was integrated into the countryside, and even accidents could be visualized in terms of the picturesque, which became "the dominant form of representing modernity" in Britain.[19] The sublime was not absent, but it was softened. As late as 2018, Jon Agar noted "the general history of the environmental and technological 'modern' cultural imaginary in Britain has yet to be written." He suggestively asked, "Has there been such a thing as a British envirotechnological sublime? If not, why not?"[20]

In contrast, during the nineteenth century, technologies that were considered sublime became a central part of American identity and helped to unite an often-divided people. The United States has been riven with political factions that were just as contentious during the presidencies of Thomas Jefferson, Andrew Jackson, or Andrew Johnson as they are today. But the natural and technological sublimes united Americans who otherwise seemed irreconcilable. Most Americans learned to identify the nation with Niagara Falls, the Natural Bridge of Virginia, Yellowstone Park, and other natural sites. They also celebrated canals, railroads, and bridges as national symbols. Appreciation of the technological sublime was not uniquely American, but it was more central to national identity in the United States than in Denmark, Spain, or Britain.[21] Indeed, new nations may have taken up the sublime as a central part of national identity more readily than countries formed before the industrial revolution.

In Canada, Roberta Stryan and Robert R. Taylor have argued for the importance of the technological sublime, as exemplified by the construction of the Welland Canal between 1913 and 1932.[22] One of the widest and deepest canals in the world, it allows ocean-going ships to circumvent Niagara Falls and reach Cleveland, Windsor, Detroit, Chicago, Thunder Bay, and Duluth. They argue that the Welland Canal has become for tourists a "monument to Canadian enterprise and achievement." Anastasia Rodgers has made a similar case for the construction of the Prince Edward Viaduct (an imperial name), which today is more commonly called the Bloor Street Viaduct. Completed in 1919, it was memorialized by the Toronto city photographer Arthur Goss. Another celebrated engineering feat illustrates the Canadian version of the technological sublime. The Victoria Bridge, built for the Grand Trunk Railway between 1854 and 1859, was then the world's longest railway bridge. The Canadian Museum of History notes that in the mid-nineteenth century, "The premier engineering achievement on the Canadian side of the Atlantic was the Victoria Bridge—a series of iron tubes resting on imposing piers—2.7 kilometers in length and completed by an army of some 3,000 workers in 1859."[23] William Notman's images of its construction featured the workers as well as the heroic scale of construction. His photographs were widely distributed in books and prints, made into popular lithographs, and are still exhibited.[24] Both the naming of the bridge after the British Queen and its dedication in 1860 by the Prince of Wales suggest that during the nineteenth century the Canadian technological sublime was subsumed by the British Empire's projection of its power through railways, bridges, buildings, and electrical stations constructed around the world.[25]

Canadian identity became more independent during the twentieth century. Especially after World War II, it evolved beyond the disintegrating British Empire, to emphasize Canada's sheer size and the technological projects that made it possible to overcome the vast distances and weld the nation together. Blair Rebecca Stein demonstrates how these themes recurred in the publicity of Trans-Canada Airlines, "a state enterprise designed to forward collective visions of nature, technology, and nation through the implementation of largescale infrastructure."[26] She argues,

"The Canadian sublime retained more of its focus on the natural world than the American," but still embraced "the reconstruction of the life-world by machinery" and reimagined the "dislocations and perceptual disorientations caused by this reconstruction in terms of awe and wonder." The "Canadian culture of bigness was supported and dismantled in the same breath, turning the disorienting effect of time-space compression and seemingly unnatural altitude into an awe-inspiring Canadian experience of nature."[27]

A related pattern seems evident in India. At first, as Pramod K. Nayar shows, the English promoted an "imperial sublime." Between 1759 and 1820, British travel writers adopted "the aesthetics of the sublime, with its emphasis on emptiness and waste" in India, which prepared "the ground for colonial representations of practices of a georgic Indian landscape of renewal."[28] This renewal began with agriculture,[29] and continued through technological improvements, including better roads, railroads, irrigation systems, the telegraph, a postal system, and newspapers. In short, the British saw themselves transforming India in much the same way that they were developing Britain itself. The technological changes were particularly striking to the colonized population. "Trains in India replaced gods. Nirad C. Chaudhuri writes in his *Autobiography* of the awe they produced, even in well-to-do middle-class families. . . . People bowing to passing trains, as though they were embodiments of Juggernaut, was a commonplace sight."[30] Both Canada and India retained aspects of British culture, but after independence evolved new identities. As these observations suggest, it would be worthwhile to study to what extent forms of the sublime became part of European and American imperialism.

During the twentieth century, nations that underwent rapid modernization also celebrated large engineering projects as technologically sublime. The Soviet Union applauded an enormous construction project, the Dnieper Hydroelectric Station, which on completion in 1932 was the largest dam and hydroelectric plant in Europe (figure 8.1). It also raised the level of the Dnieper River by 110 feet, with a canal and lock system that established a water route from the Baltic to the Black Sea. The electricity generated made possible the new industrial city of Sotsgorod, perceived at the time as a socialist utopia. Thousands of tourists came, and their responses, Nicholas Kupensky argues, exemplified "the Soviet industrial

8.1 The Dnieper Dam in Ukraine, almost one kilometer long, was the largest hydro-electric dam in Europe when completed in 1932. One of Joseph Stalin's most successful projects, it raised the Dnieper River by more than 110 feet, rendering it navigable for large ships. Max Alpert, photographer. Courtesy of Pinterest.

sublime."[31] He argues that this industrial sublime differs from that in the United States because nineteenth-century Russia had a more ambivalent attitude toward factories and large technical projects. After the Revolution, the Bolsheviks "inherited a crippled industrial infrastructure" and even lacked a close equivalent to the word "sublime" in translation, and essentially had to invent this tradition.[32] Susan Buck-Morss suggests in *Dreamworld and Catastrophe* that the "Soviet sublime" emerged in the effort to imagine and build enormous projects, by "overcoming the physical limits of the collective workers" whose individual diminishment in the collective made the enormous dam possible.[33] Unlike most examples of the industrial sublime, the Dnieper Hydroelectric Station was intentionally destroyed. When forced to retreat before the advancing German armies during World War II, the Soviets did not want the Nazis to acquire the canal system and hydroelectric station. Instead, they were made inoperable as part of a scorched earth policy.

During the 1930s Germany had its own version of the technological sublime, notably as expressed in the architecture of Albert Speer, who designed impressive structures for the Nazi regime. Robert Jay Lifton noted, "So grandiose were the projections he and Hitler made together that some of the buildings were to hold as many as 150,000 people on vast balconies in a new Berlin that would become the center of the world, dwarfing the grandeur of Paris and the Champs-Elysees. Few of the structures were actually built but many were imagined."[34] The technological sublime also found expression in the hegemonic displays adopted by Mussolini in Italy. In contrast, such hortatory architecture is unusual in egalitarian nations such as Denmark or Norway. They can afford to build skyscrapers and impressive public buildings but generally have chosen not to do so.

China had a different context for its version of the technological sublime. Brady notes that the Chinese did not have a tradition of "the subject/object dualism of Kant and other Western thinkers."[35] As Matteo Tarantino explains, during most of Chinese history the attitude toward machines was primarily practical, and they were not understood in metaphysical terms. Unlike the West, the Chinese did not conceive human destiny in terms of a millennia-long struggle to recover from a fall from Eden. But their view of technology changed after the "Century of Humiliation" when China was repeatedly defeated by foreign powers, beginning with the Opium War of 1839, and culminating in the Japanese occupation during World War II. In response, twentieth-century Chinese intellectuals called for scientific and technical education, with the goal of recovering the country's power and self-determination. The machine became sublime not so much as a vehicle for personal transcendence or as a means to dominate nature but rather as a tool to revive and reconstruct the nation. By mastering technology, China could escape victimization by foreign powers and again become a leading nation. In that struggle, Chinese scientists became unequivocal heroes. In contrast, popular European and American narratives were more likely to depict diabolical scientists and the unanticipated consequences of science and invention. The Chinese saw the technological sublime in terms of national rejuvenation, symbolized by its space program, achievements in genetics, and engineering projects such as the Three Rivers Dam, which in 2012 was the world's largest hydroelectric plant. Although this project met resistance as it submerged 1,200 historic sites and displaced 1.3 million people, nevertheless, most Chinese retain "an overwhelmingly positive attitude toward technology," which also is expressed in their predilection for skyscrapers.[36]

Argentina has also embraced the technological sublime. Camila Costa examined three gargantuan construction projects traversing the Paraná River, the second largest in South America, which separates provinces north of Buenos Aires from the rest of the country.[37] First came the three-kilometer-long Uranga–Sylvestre Begnis tunnel under the river, built between 1962 and 1969. Next, a 1,700-meter suspension bridge, the General Manuel Belgrano Bridge, was built between 1967 and 1973 (figure 8.2). It united the cities on either side into a single conurbation and provided for both pedestrian and automobile traffic thirty-five meters above the river. Lastly, near Buenos Aires the Zárate–Brazo Largo railway bridge was built between 1971 and 1979. These works were celebrated in elaborate opening ceremonies, honored on postage stamps, depicted on postcards, and given extensive press coverage. Symbols of local and national identity, they can still astonish visitors approaching the river, as the enormous scale of construction becomes apparent and dramatic vistas unfold. The three projects demonstrate how technology can reconstruct space,

8.2 The 1,700-meter-long General Manuel Belgrano Bridge over the River Paraná was completed in 1973 and has become a symbol of Argentina's modernization. Courtesy of Wikipedia, https://en.wikipedia.org/wiki/General_Belgrano_Bridge.

unite divided sites, and materialize the state's power, becoming emblems of modernization. In recent years, however, a prolonged drought has reduced the flow of the Paraná by more than half, harming fishing, shipping, and water supplies, and making it an example of the devastated environmental sublime.[38]

Other national examples could be added, including the French high-speed trains and nuclear power plants that have become emblems of technological prowess and modernity.[39] Yet more examples are not necessary to demonstrate that while sublime experiences appear to be universal, their interpretation varies based on national history, geography, technological capacity, and cultural values. A similar argument might be made concerning the disastrous sublime, for each nation has experienced different types of catastrophes, and some societies obviously endure more of them than others. The Japanese have frequent earthquakes, Spain has few. Hurricanes recur every autumn in the Caribbean, but they are almost unknown in Sweden. Likewise, the martial sublime varies in importance and interpretation from one nation to another. The natural, technological, disastrous, and martial sublimes can all be woven into a national identity, but each country selects and interprets differently.

Yet not all forms of the sublime lend themselves equally well to nationalism. The disastrous sublime may unite nations more often than it divides them, for fires, earthquakes, tsunamis, pandemics, and floods pay little attention to borders. The tsunami of 2004 struck the islands and coastlines of many nations in the Indian Ocean, killing more than a quarter of a million people. In the aftermath, the countries affected developed an international warning system to minimize the loss of life in future disasters. The horrific fires that raged out of control in Australia in 2020 dispersed smoke into the upper atmosphere that circled the globe, affecting all other nations in the southern hemisphere. Disasters often remind people that they are interconnected and ultimately share the same fate.

Scientific research also tends toward international cooperation. A nation may be proud of particular scientists, but systems of scientific knowledge do not belong to any country. Indeed, high costs often push nations to pool resources. Europeans collectively have one space program and one giant collider at CERN. Despite political tensions, the Americans and Russians cooperatively maintain a space station and rocket launch site. And the famous image of a gigantic black hole publicized in 2019 was only possible because eight of the largest telescopes in the world cooperated to make simultaneous observations. No single nation could take credit for the result. On a smaller scale, the members of research groups often come from different nations. A discovery may be made in British laboratory by an international team. People from different nations often share a Nobel Prize.

Likewise, scientists from many nations work together to identify new species; to understand the migration patterns of birds, fish, insects, and mammals; to measure global warming, pollution, and species extinction; and to study other matters central to the environmental sublime. The intimate knowledge of the local required by the environmental sublime does not lend itself to a new form of nationalism, for the growth or collapse of any habitat has ripple effects that extend beyond national borders. Birds annually migrate between Canada and Mexico, or between Europe and Africa, and the extinction of any one of them can have damaging effects along the entire path of their annual journeys. In the past, nations have been prone to interpret spectacular landscapes and technologies as monuments to national greatness, but the environmental sublime recognizes the artificiality of national borders. The ecology scarcely changes when one crosses from North Dakota to Canada, from the Amazon basin in Columbia to the same basin in Brazil, or from the southern Netherlands to the Flemish section of Belgium. Nationalism is not a meaningful category when studying currents of air like the Jetstream. Neither animal migrations nor air pollution nor the weather stop at the frontier. The technological sublime is usually linked to finite spaces. It is a cornerstone of the older nationalism that emphasizes borders, walls, tariffs, and exceptionalism. In contrast, the intangible and environmental sublimes entail a global view and international cooperation. Most environmental crises demand not a national but an international response.

Taking a broad view, the seven forms of the sublime have had different relations to nationalism. The natural sublime is compatible with conceptions of an ancient or even eternal relationship with a specific location. Such ideas often were part of national identity before the revival of the sublime in the early modern period. As was the case with Denmark, Britain, or Spain, the natural sublime competed with preexisting focal points of nationalism such as a royal family, ancient ruins, famous battle sites, and so forth. The natural sublime is often prominent in newer nations, such as the United States, Australia, and Canada, where the older trappings of nationalism were less important or even absent. Most examples of the technological sublime emerged after 1750, being chiefly the machines and systems developed during the industrial revolution such as factories, railroads, steamships, canals, dams, skyscrapers, airplanes, and space programs. Nations that rose rapidly to international prominence based on such achievements, particularly nations that experienced political revolutions that transformed their political systems, often embrace this form of the sublime. This is strikingly the case with the United States, for example. Nations seeking to celebrate a newly achieved modernity, such as Argentina, may also emphasize the technological sublime.

The seven sublimes often contradict rather than reinforce one another. Identification of Niagara Falls and the Grand Canyon with the natural sublime conflicted with the full development of hydroelectric power at both sites. Likewise, the knowledge and sensibility developed through the environmental sublime usually increases awareness of pollution, waste, species extinction, and global warming, which can lead to conflict with the monumental highway and building projects celebrated in the technological sublime. More generally, it appears that the material sublimes have been more easily subsumed within nationalism than the mediated sublimes, which tend to foster a more international outlook.

Each nation has a different relationship to the seven forms of the sublime. Some emphasize landscapes, while others develop a preference for the more spectacular, dynamic forms of the sublime. While there are many possible combinations and interpretations, the historical pattern appears to have three main aspects: the movement from a few to many sublimes; a shift from nationalism to internationalism; and increasingly conflicted national identities. This important subject becomes more accessible to research once the variety and differences between the seven sublimes are understood. The following chapter compares these seven forms of sublimity, each of which has a different relationship to space and time.

9

SUBLIME LANDSCAPES AND SPECTACLES

This study of the different forms of the sublime has focused on historical examples. It took the eighteenth century as its starting point, when Edmund Burke and Immanuel Kant both were concerned primarily with the natural sublime. During the following three centuries, this category of experience has expanded to include constructed landscapes, disasters, battles, scientific observations, digital practices, and environmental complexity. Most philosophers have been reluctant to accept the public's usage of the term "sublime" to describe experiences that seem incompatible with Kant's philosophy, where sublimity has a central role in the development of Reason. And yet the possibility of heterogeneous sublimes was already hinted at in the writings of Burke and Kant themselves, which briefly addressed some aspects of architecture, disasters, warfare, and the discovery of new sublimities in the universe.

Chapter 1 began with Dickens's visit to Niagara Falls and his ascent Mount Vesuvius, as illustrations of sublime experiences that accord well with Burke's and Kant's descriptions of the natural sublime. It then reviewed Kant's division of the sublime into the mathematical and the dynamic forms.[1] The mathematical sublime is epitomized by Wordsworth's encounter with Mount Blanc and the dynamic sublime by Dickens climbing Mount Vesuvius. This fundamental division suggested two modes that were re-expressed in other sublimes, each of which can be experienced as either an immense landscape or as a dynamic spectacle. Sublimes based upon direct sensory experience are the sort that Burke had in mind, where many senses are deeply engaged. For example, amid battle one feels and hears explosions, smells and tastes smoke, and automatically combines these sensations with a view of the battle, to create a single overwhelming impression. An assault on all the senses also occurs in the quieter encounter with the Grand Canyon. One visitor recalled, "What the canyon does for you is to pick you out of circumstances and scenes and away from people and ways of living that do little to stimulate your awe and reverence, and to set you down in the place where, try as you will to resist, try as you will to hold yourself in another mood, you cannot help yielding to your awe." In addition to the powerful colors, the uncanny sense of space, and the unavoidable contrast between human time and the vast time of the Grand Canyon, this visitor felt the powerful effect of its great maw of silence: "that hush of eternity that broods over it day and night is one of the most mysterious and one of the most powerful things with which I personally have ever come into contact."[2] The classic forms of the natural sublime are direct, physical encounters whether experiences of landscapes like the Grand Canyon or of spectacles like Mount Vesuvius in eruption.

Equally direct encounters with newer sublimes were the subjects of chapters 2–4. Chapter 2 examined the technological sublime that emerged in the eighteenth century, shortly after the natural sublime, expressing appreciation for spectacular demonstrations, large-scale engineering projects, and impressive architecture. Burke partially anticipated it in a few remarks on architecture. As with Kant's delineation of the natural sublime, it has two modes. One is a response to the static immensity of technological objects such as bridges, dams, and skyscrapers; the other is a response to dynamic, moving objects such as hot-air balloons, railroads, automobiles, airplanes, and rockets. Some sites are both landscapes and spectacles, such as the Panama Canal or a vast factory complex. The substitution of human constructions for nature was a fundamental change. In the nineteenth century some commentators argued that the technological sublime had a moral dimension. Hepburn and others still argue that the natural sublime retains a metaphysical dimension, but few now make such a case for the technological sublime. Nevertheless, impressive human creations can evoke awe and may lead to reflections about the role of humanity on Earth and the ultimate trajectory of history.

This trajectory is not necessarily a story of progress. Chapter 3 turned to disasters, which Burke discussed, noting its two modes, either as a catastrophe that is witnessed (a spectacle) or as a retrospective view of past devastation (a landscape). Disasters have both natural and technological causes. For example, San Francisco is built on a major geological fault that guarantees it will suffer repeated earthquakes. Not all who live through disasters experience them as a sublime spectacle. For example, that is inconceivable for people trapped inside elevators or beneath the rubble of a collapsed building. However, a disaster may be experienced as sublime by those who observe it in safety, as was the case for some during the 1906 San Francisco earthquake and fire or the Great Chicago Fire of 1871. In the spectacular form of the disastrous sublime the sense of safety diminishes compared to the technological sublime, because human beings have lost control. Where the technological sublime is exhilarating and focuses attention on the self, in the disastrous sublime one is also absorbed in the fate of others. The landscapes of historic disasters such as Pompeii are static, melancholy ruins. Some can be seen only for a short period before rebuilding. Other ruins have endured and been enshrined in popular culture.

Burke's remarks on the effects of artillery fire on the body to some degree anticipated the martial sublime, which like the technological and disastrous sublimes also has two modes. The landscape mode could be a view of opposed armies seen in the distance from a hillside before a battle begins. But in most cases, it is a retrospective view of a battlefield, or a panorama representing the event. Major battles such as Waterloo and Gettysburg were among the most popular subjects of panoramas. In contrast, the dynamic form of the martial sublime can only be fully known in battle, but not all troops have this experience. It requires an expansive view of battle from a site that is at least temporarily safe. It is terrifying and potentially traumatic.

The four sublimes examined in the first four chapters share important characteristics. They do not require prior knowledge or expertise, and they are accessible to anyone. A hallmark of the sublime is that it is universal, and not limited to one gender, one class, one racial group, or an elite. These experiences all may arise suddenly without preparation, and their impact may be all the greater when they are unexpected. People who acquire a good deal of information about a famous waterfall, sky-scraper, or other site may, when they see it, have difficulty experiencing it as sublime because their expectations interfere with their perceptions. These unmediated sublimes are experienced directly through the senses, and no specialized equipment is needed. Horror, in the sense that Burke defines it, intensifies these experiences and focuses the mind. All of these sublimes can be photographed, filmed, or otherwise made into represen-tations. The Panama Canal was often photographed and filmed, and it was even reproduced as a gigantic scale model at the 1915 San Francisco World's Fair.[3] These representations may be powerful, but they do not engage all the senses, and they are not a substitute for direct sublime experience.

Each of the directly experienced sublimes is divided into landscapes and spectacles. The landscapes are quiet or even silent, for example the view of a large city from the top of a skyscraper, where the sounds from below are faint; or when visiting the ruins of Pompeii. They provide impressive, static, safe experiences of space, often seen from an elevated position. Landscapes such as the Gettysburg Battlefield change little over time, and they can be visited repeatedly. With assistance from gardeners and preservation experts, sublime landscapes may remain the same for generations. Some do disappear, however, when a site is reused or rebuilt, as was the case with the Chicago Columbian Exposition after it closed in 1894. Its ruins could only be viewed for a few months. These landscapes may, like a view of the Milky Way, lead to thoughts that go beyond what is immediately visible. Sublime landscapes can seem to exalt humanity's achievements, but they also may lead to thoughts of life's ephemerality, of how quickly a prosperous city can become a vast ruin, or of how weak the individual is compared to modern weaponry. The sublime landscapes suggest both greater human powers and increasingly destructive forces. They are static, yet they intimate wrenching change. They invite medita-tion on the universe, fate, and other large questions.

In contrast, sublime spectacles are ephemeral, dramatic, and unique, and they do not usually provide much leisure for speculation. Volcanic eruptions, earthquakes, rocket launches, and battles are loud, active, and

immediately dangerous. There is a decisive moment. To experience a sublime spectacle, one must be near the volcano when it erupts, on the coast
when the hurricane hits, in the trenches during a firefight, or in the airplane when it flies through a lightning storm. One stands at the edge or
outside of sublime landscapes and gazes at them; but one is immersed in
sublime spectacles, and they have a visceral impact. Given their transi-
tory nature, they are less commonly experienced than sublime landscapes,
which are more readily accessible. One can plan a visit to a mountain, a
skyscraper, or a historic battlefield more readily than one can schedule
travel to a hurricane, earthquake, fire, or battle. Some dynamic sublime
experiences are preserved and become static memorials. Pompeii was a
dynamic sublime event for the few who could see it in safety two thousand
years ago, but since its excavation in the middle of the eighteenth century,
Pompeii has become a quintessential landscape of the disastrous sublime.
Dynamic spectacles may be preserved as landscapes or simulated in panoramas, amusement park reenactments, and films. However, these representations have less impact than being personally caught up in the events.

Each form of the sublime implies a different spatiality. The space of
the natural sublime is beyond human control and signifies the weakness
of humanity before the immensity of nature. In contrast, in the technological sublime human beings seem to master space and subject it to
their control. Canals connect the Mediterranean and the Red Sea or the
Atlantic and Pacific Oceans; great bridges span the Mississippi, the Rhine,
the Bosporus, or the Danube. Drilling for oil and mining subject nature to
humanity's needs, for example at the world's largest lithium mine at Salar
de Atacama, in Chile.[4] Its enormous evaporation pools, where lithium is
separated from other substances and concentrated before shipment to
refineries, are visible from outer space. The technological sublime can
be abandoned, erased, or reconstructed. Dams can be dynamited or skyscrapers torn down so that even larger structures can replace them. The
dynamic, more immersive forms of the technological sublime express
another relation to space, as it is conquered through rapid movement,
often to the point where the landscape becomes a blur, an almost irrel-
evant background.

Disastrous sublime landscapes are static scenes of death and destruction, such as New Orleans after Katrina, Chernobyl after the nuclear

meltdown, or anywhere that human beings have failed to control or resist powerful forces. In contrast. the dynamic mode of the disastrous sublime is only possible when sheltered from these forces in a safe space. There, the observer is immobilized. Finally, the two modes of the martial sublime present battle as both landscape and spectacle. The landscape mode is not perilous, and it is primarily visual, though other senses are involved. Some sites are impressive in themselves, such as the cliffs in Normandy overlooking the beaches of D-Day, where thousands of troops came ashore under withering fire, drove back the German army, and began the liberation of France. Yet often a battlefield is a large open space or some rolling hills that in and of themselves are hardly sublime but become so when invested with memory. In contrast, immersion in the spectacle of battle may be experienced more sonically than visually. One can hear but not see gunshots, bombardments, and incoming shells, and one usually hides in the face of an attack. The space of survival is reduced to a trench, foxhole, or space behind a wall or rock. If these refuges afford enough safety, then a sublime experience may be possible. Later, when the battle is over, the place where it occurred is redefined as a landscape of war, and its meaning is measured by human sacrifice. The villages, fields, hills, and woods are redefined in terms of tactics, attacks, defensive positions, and the like. Particular locations are renamed in terms that suggest heroism, tragedy, sacrifice, and slaughter, and parts of the scene are often requisitioned as a cemetery. The land has lost its economic use value and become a symbolic site that marks victory and defeat, forever linked to the time when the battle raged.

Each sublime implies a different temporality. The time of Kant's mathematical sublime seems eternal, measured in the billions of years revealed by geology or in the light-years required to traverse the Milky Way. In the nineteenth century, it was often associated with God. In contrast, the technological sublime concerns only a narrow window of recent human history, at most a few thousand years, usually far less, and it is associated with human achievements. Recently constructed dams, skyscrapers, and other productions are taken to be proof of progress. In the technological sublime accelerated travel compresses time, in an apparent rush toward the future, in ever faster railroads, cars, airplanes, and rockets. In contrast, the disastrous sublime is not about human purposes or powers of transformation but about human weakness in the face of fire, flood, earthquake, tornado, or volcano. Catastrophe traces no line of progress but breaks time into incoherent pieces and obliterates the future. It points at the moment when powerful forces struck and condemned many to an inscrutable fate. It rivets terrified attention on unexpected and uneven rhythms, like the winds of a hurricane or an earthquake's unpredictable aftershocks. There is no order or purpose in catastrophic time. The landscapes of the disastrous sublime preserve a frozen tribute to random destruction: the lost city, the burst dam, the immense wreckage. If disasters seem purposeless, war is about conflicting purposes. The martial sublime concerns epic military events. Its witnesses are immersed in a violent spectacle, with an abrupt back and forth between stasis and savage action, caused not by natural forces but by implacably opposed troops. A hurricane has no purposes, but a battle has many, including self-preservation, revenge, defense, and conquest. Death in the disastrous sublime is an accident; in the martial sublime it is random and yet intentional. In both, time comes to a violent halt. For those who escape death, these moments are like fragments of shrapnel, disconnected from the rest of life. As Gray recalled, "So often in the war I felt an utter dissociation from what had gone before in my life; since then, I have experienced an absence of continuity between those years and what I have become."[5]

While these four tangible sublimes imply different conceptions of space and time, they are experienced directly with all the senses. In contrast, the intangible sublimes have quite different characteristics. They are mediated by equipment, including microscopes, telescopes, satellites, computers, and drones. Preparation, which can be an impediment to direct sublime experience, is unavoidable and essential to mediated sublimes, both to gather data and to interpret it. Where unmediated sublimes are individual, often spontaneous experiences, most mediated sublimes are constructed experiences that require teams working together who share the results. Mediated sublimes must be constructed before they can even be sensed. While photographs, films, or sound recordings can be made of the directly sensed sublimes, they are not necessary to have those experiences. In fact, images of the tangible sublimes are usually denigrated as being inadequate representations compared to the powerful experience of being there. In mediated sublimes, however, there is no direct sensory perception, and representations—of a distant galaxy, the surface of Mars, the structure of DNA, or a subatomic particle—are the basis of the experience. The mediated experience is seldom dangerous. Unlike Dickens on Mount Vesuvius, who was showered with sparks while gazing down at molten lava, the mediated sublimes typically involve looking at a computer screen, photograph, or data. The element of horror neither stimulates nor intensifies these experiences, which from beginning to end are more cerebral than any of the tangible sublimes. Moreover, in mediated sublimes, the sense of time is transformed because experiences are first recorded, then constructed, and finally played back. In many cases, like the images from the Hubble Space Telescope, the data has been modified in the process of reconstruction. This does not make these experiences inauthentic, but they always come after the fact, due to the transmission time from a rover or satellite to Earth and to the need to process data before it can be interpreted.

This shift from tangible to intangible sublime experiences that are only accessible through mediating technologies began to emerge by the time of Galileo's telescope. It was already common in the late eighteenth-century biology, astronomy, and physics. Since that time a plethora of instruments have been invented that reveal intangible objects, qualities, and relationships. One might be tempted to lament this change as a displacement of attention from direct contact with the physical world to abstraction, but this would be misguided, for there is no other way to learn about the far reaches of the universe, microscopic life, or subatomic particles.

The sublime produced by a team of scientists is experienced as a region, either somewhere else in the universe or a microscopic realm invisible to the unaided senses. The knowledge of what we cannot know through the unaided senses has increased rapidly in the last twenty-five years. More powerful instruments have revealed planets outside the solar system, shown that the Milky Way contains more stars than previously thought, and produced images of black holes. These wonders can only be witnessed indirectly as data and images produced after the observations. Yet this does not necessarily lead to scholarly detachment. The experiences of scientists and oceanographers driving rovers demonstrate a strong identification with the technological apparatus that is a surrogate for human presence. Not every new discovery is sublime, but human understanding now includes myriad invisible objects and life forms. As

Diane Ackerman put it, "Our mental cosmos teems with a thicker texture of invisibles than ever before."[6] We understand them through scientific diagrams and images from microscopes, telescopes, space probes, and sensors of many kinds.

These technologies have been linked to computers. Yet not all digital technologies are sublime, any more than all tall buildings are. Some landscapes of digital hardware are impressive, notably in the largest server farms, and they often are presented in ways similar to the classic images of vast factory interiors. The most promising digital sublime to date is virtual reality, which emphasizes movement and shares some characteristics of other sublime spectacles. However, VR's dramatic experiences are often scripted and goal oriented. VR differs from the intangible sublime, which uses instrumentation to discover objects inaccessible to the unaided senses. But digital experience has a more tenuous connection to external reality. Gaming, VR, and Internet-based software can simulate objects, often as part of dramatic, immersive fantasies that invite intense identification. VR has the potential to provide new kinds of sublime experience, not so much by simulating the world as human beings already know it, but rather by showing what it would be like to have a different body and a new sensorium, like that of a bird, an octopus, or another companion species.

There are other distinctions between the intangible and virtual sublimes. The intangible object may be sublime not only because of its size, power, or other characteristics but also because it has scientific interest. Did Mars have rivers? Did it once support life? Mediating technologies make it possible to answer such questions. In contrast, a computer game can be mesmerizing and intense, but it seeks answers to no large questions, and beyond the program there are uncertain (or nonexistent) physical reference points. Virtual reality makes possible artificial experiences, such as visiting a simulated Grand Canyon. In a suspension of disbelief, the mind can learn to accept virtual stimuli as real. The digital sublime exalts not God, nature, the nation, engineers, or scientists, but the technological self. It emerges not from contemplation of nature, nor from an impressive physical construction, disaster, or battle but from software. The mental process involved is not the transcendental Reason of Kant contemplating nature, nor the triumph of rationality embodied in a new dam, but rather the play of the mind when using an interactive program.

In contrast, the environmental sublime is only partially based on instruments that provide otherwise invisible information about a place. They answer questions such as: What is the precise chemical composition of the soil? Is radioactivity present? Are microplastics in the food chain? But the environmental sublime is a hybrid form that is also based on long-term immersion in a place, and it requires direct observation. It does not arise from an encounter with that which is absolutely great but rather with ecological systems that are infinitely complex. It focuses less on spectacular sites or dramatic events than on complex, wondrous, and often endangered ecologies. These evoke awe the more they are studied and contemplated. The environmental sublime has two modes. The first is a noninvasive encounter, seeking to understand ecological connections and symbiotic relationships. Henry David Thoreau might be considered an early practitioner; Rachel Carson and Jane Goodall are role models. In the second mode, an observer witnesses environmental destruction, possibly including species extinction. This experience resembles the disastrous sublime, but compared to the destruction of cities by earthquake, fire, or flood, it is more absolute. A city can be rebuilt, but the carrier pigeon and the Tasmanian tiger are gone forever, along with their role in environments they once inhabited. The witness is forced to meditate not only on the ephemerality of a particular life but also on the survival of species, ecosystems, and humanity itself. This violence is in its way as horrifying as what a soldier sees in battle, and no less painful for being in slow motion. The environmental sublime evokes a different conception of time, far longer than the short-sighted hubris of the technological sublime but shorter than the billions of years of the intangible sublime. Robert Macfarlane suggests that environmental awareness is based on "deep time," which "is measured in units that humble the human instant: epochs and eons, instead of minutes and years. Deep time is kept by stone, ice, stalactites, seabed sediments, and the drift of tectonic plates." Deep time began long before humanity existed. It prompts us to "see ourselves as part of a web of gift, inheritance and legacy stretching over millions of years past and millions to come." In "deep time, things come alive that seemed inert." Ice caps move. "Rock has tides. Mountains ebb and flow."[7] At the same time, awareness of global warming, species extinction, and other environmental changes can induce a nostalgia for the apparent constancy of an earlier time, for example when snow fell every winter, the summers were not so scorching, the hurricanes less frequent, or the fish more abundant in the oceans.[8] The environmental sublime increases sensitivity to change and explores a much longer chronology than the brief human histories that are the implicit frameworks of the technological, disastrous, or martial sublimes. Yet the time of the environmental sublime is far shorter than the eternities of an astronomer surveying the universe.

There is a sharp distinction between the sublimes analyzed in chapters 1–4, which can be experienced directly with the senses, and the sublimes examined in chapters 5–7 that are largely or entirely experienced through mediating technologies. Wherever the sublime is encountered directly, it includes an element of danger and a sense of horror. But there is no danger when viewing satellite images or looking through a powerful microscope, and such mediated sublimes do not engage the full sensorium. Moreover, these mediated experiences typically demand expertise in using complex equipment, combined with specialized knowledge. Such sublimes must be contextualized and interpreted for outsiders who lack these qualifications. To put this another way, the mediated sublimes do not make as strong an impression if the witness has not acquired expertise beforehand. The differences between the spectacle and the landscape forms of mediated sublimes are less pronounced than with material sublimes. These generalizations do not fully apply to the environmental sublime, however, as it combines experiences mediated by scientific instruments with close, direct observation of nature. It is a hybrid reformulation of the natural sublime.

There is not a single sublime that varies slightly depending on the object or site encountered. Rather, there are seven distinct *formations* of the sublime, each with different spatial, temporal, and sensory frameworks. For example, both flying a fighter jet and directing a drone require expertise, but only the fighter pilot has a direct experience of flight that demands a heightened engagement of all the senses. The pilot is in danger; the drone operator is not. Only the pilot is in a potentially terrifying situation, and therefore is in a position (potentially) to experience the sublime. The experience of drone operators more closely resembles that of a NASA scientist who by remote control drives a rover across the surface of Mars. Both need expertise; both are in completely safe locations; and both have an experience entirely mediated through technology. Even so, there is a fundamental difference. The drone operator's objective is to kill people, whom he or she first identifies and observes. The rover driver explores an unknown world, looking for new scientific information. That objective is potentially sublime; the drone operator's is not. This leads one to reconsider the fighter pilot, whose ultimate objective is no more sublime than the drone operator's. If the fighter pilot experiences the sublime, it is not because he or she kills, but because of the powerful sensations of flight itself, sharpened by a sense of danger.

Examination of the intangible and environmental sublimes strongly suggests that the distinction between the natural and the technological sublime has become untenable. Nature and culture have become intertwined, whether one considers a sunset made more colorful by industrial pollution, an inundation caused not only by a hurricane but also by building on a flood plain, or the impossibility, in practice, of keeping a wilderness area free of smog, acid rain, invasive species, and human intrusion. The collapse of the dichotomy between Nature and Culture should not be understood as a loss, however, as though nature once was untouched and now has "fallen." Human beings have never been separated from nature.

Yet if the distinction between nature and culture has collapsed in most academic disciplines,[9] both the natural and technological sublimes persist as experiences that heighten our engagement with the world. As Brady emphasizes, it would be a mistake for modern philosophy to repudiate the natural sublime on the grounds that it is pantheistic or metaphysical. Rather, she argues, the sublime enables "a more pluralistic conception of the aesthetic valuation of nature."[10] Might not a similar view be adopted toward the technological, disastrous, and martial sublimes, each of which entails a different conception of humanity's place in the world? Moreover, the environmental sublime builds upon the natural sublime, with the important difference that it is grounded in mediated scientific observations and close study of local ecologies. In contrast to how the natural world was perceived in Kant's era, the environmental sublime inescapably confronts the dangers of the Anthropocene in a world that is no longer safe. One may feel, as Aldo Leopold put it, that one is "alone in a world of wounds."[11] Like the experience of battle, whose intensity can border on trauma, the sense of blight can become an incapacitating vision of apocalypse and contribute to a sense of irretrievable loss, and even paralysis in the presence of apparently inexorable and rapid decline. It can become the absolute contradiction of the progressive vision and wonder of the technological sublime.

It is no accident that the history of the sublime since the eighteenth century coincides with industrialization, scientific discovery, and the development of increasingly complex technologies. But the relationship is not a simple one of cause and effect, as can be seen in Kant's discussion of the sublime. Key examples for his argument came from astronomical discoveries such as the vast extent of the Milky Way. Yet he does not present the sublime as a mere side-effect of scientific investigation. Rather, the sublime begins as a powerful sensory experience of wonder that anyone might have when encountering overwhelming forces or objects, which in turn leads to reflections about the meaning of such encounters. Science does not simply produce such experiences; rather the experiences stimulate and reinforce curiosity, including scientific study. The sublime is not simply a stimulus or a response, but part of a self-reinforcing process that leads to discoveries and to an increasing variety of sublime experiences. These have grown to include technological sublimes, their inversion in destructive and martial sublimes, and their extension into intangible and virtual sublimes.

In all these areas, technologies have become more important, reaching a point where mediating devices and statistical measures increasingly are an inseparable part of sublime experiences. Hurricanes are understood in terms of wind speed, storm surges, and size, and presented as radar images, maps, and photographs. Earthquakes are evaluated on the Richter Scale. Not only are ever more aspects of sensory experience registered and evaluated through mediating devices, but many new experiences can only be known through the filter of machines. At the same time, some earlier formations of the sublime no longer impress. Steam railroads, once thought sublime, now seen quaint, and airplanes have become almost banal. The formations of the sublime are stable, but their contents change and the number of possible sublime experiences increases. While their individual intensity may become attenuated, it is not always for the same reason. Some forms of the technological sublime pale as they become familiar; once-famous battles fade in importance for later generations; massive natural disasters that occurred more than a lifetime ago may be all but forgotten; and experiences mediated by technology may become banal by repetition. Yet taken together, the many forms of the sublime are a repertoire of powerful experiences, that become (often conflicting) parts of a nation's or an individual's self-conception. The resulting identity selects from and combines spectacular scenes, technological achievements, disasters, battles, scientific discoveries, virtual reality, and environmental circumstances.

The many sublimes are in tension with one another. For analytical purposes, this book has separated and compared them. But in life sublime experiences often occur together or in close proximity. They may overlap, reinforce, or contradict one another. Consider the sites encountered on a vacation visit to the Grand Canyon, an exemplary object of the natural sublime. To drive there from California, one passes through miles of desert, sees the spectacular electric lights of Las Vegas, and visits the technological marvel of Hoover Dam. On such a journey, tourists may also visit a virtual reality studio, Arches National Park, wilderness areas, the pre-Columbian ruins of Chaco Canyon, the Nevada site where atomic bombs were tested, and the waters of the Colorado River that drought has severely reduced. Visitors may also hike into the solitude of a remote desert. Gazing into the Grand Canyon or into the desert sky at night, they may sense the evanescence of human life. Kant argued that the natural sublime forces human beings to grapple with immensity and to develop a more profound self-knowledge. Today, faced with a multitude of intense experiences, ranging from ancient wonders to new sublimes, certainty remains elusive. Each sublime suggests different spatial and temporal dimensions; each intimates different narratives. Taken together, the many sublimes excite the senses, arouse expectations, suggest dangers, provoke reflection, stimulate creativity, expand awareness of change, and rock the foundations of certainty. These powerful emotions keep us humble and curious and remain fundamental to what it means to be human.

EPILOGUE: FUTURE SUBLIMES?

The title of this book is *Seven Sublimes*, not *The Seven Sublimes*, for there is little reason to suppose that there will not be more. On the material level, one can track the emergence of new *objects* of the sublime, from the hot-air balloons and panoramas of the late eighteenth century; to the railroads, bridges, skyscrapers, and electrification of the nineteenth century; to the dams, powered flight, space exploration, and computers of the twentieth century; to more recent holographs, drones, gaming, and virtual reality. This does not take into account contemporary efforts to merge human beings with machines; nor does it look at what some scientists anticipate as the Singularity or Omega Point, when artificial intelligence might take over society. Making any such list is not only an outline of what human beings have seen as sublime. The list may also be read as a history of humanity's progress, as the story of accelerating separation from the natural world, or as a warning that calls for a turn toward the environmental sublime.

If humanity is engulfed by the "Singularity," in which machines will surpass human intelligence, then the sublime might become an archaic emotion. Alternately, if human beings merge with machines to become cyborgs, they might value only a few aspects of the sublime. Would cyborgs value the inexpressible, the ineffable, or undefinable experiences? Would they be able to feel terror or fear, which Burke and Kant considered essential to the sublime? The sublime has always been an intensely human emotion. Could it take nonhuman forms?

Alternatively, the sublime could disappear. During a long hiatus from the fall of the Roman Empire until the early eighteenth century, few people were interested in the sublime. Will new forms of sublime continue to emerge, or might they begin to wane? This question seems inseparable from the question of to what degree human beings will use new technologies to modify their bodies and enhance their minds. Adding mechanical parts to human beings has gone on for centuries, including false teeth, wooden legs, and eyeglasses, and more recently hearing aids, pacemakers, and hip replacements. In 2021 scientists were experimenting with connecting computer chips to the brain. To some people the merger of humans and machines seemed imminent.[1]

The idea of creating living machines is ancient. Ingenious automata were famous in ancient Greece and Rome, and in 968 AD a traveler in Constantinople was astonished by wonders such as a "bronze tree with automata in the forms of a lion and birds of different species, which roared and chirped as people approached."[2] Centuries later the philosopher René Descartes developed "a mechanistic description of the human body," but to avoid conflict with the Catholic Church, he withheld it from publication until after his death.[3] Eighteenth-century Parisians flocked to see "a life-sized mechanical statute that could play fourteen airs" on a flute. To emphasize that the statue was not magical, its inventor Jacques de Vaucanson was pleased to reveal its springs, levers, and gears. Nineteenth-century literary works explored the question of whether mechanical men and women could be infused (often using electricity) with intelligence, passion, or a spiritual nature. Utopians speculated that robots might enable humanity to escape heavy labor, while dystopians feared that robots might make humankind superfluous.[4]

The word "cyborg" first appeared in 1960 in an article by space scientist Manfred Clynes. The mechanical elements of the cyborg were "to function without the benefit of consciousness, in order to cooperate with the body's own autonomous homeostatic controls."[5] But the potential architecture of the cyborg became more ambitious after the decoding of DNA.[6] The cyborg seemed, in Donna Haraway's words, to be "the enhanced man who could survive in extraterrestrial environments."[7] The cyborg suggested, "Escape from the earth, from the body, from the limits of merely biological evolution." She declared, "Man is his own invention; biological evolution fulfills itself in the evolution of technology."[8] In the late twentieth century, she argued that "machines have made thoroughly ambiguous the difference between natural and artificial, mind and body, self-developing and externally designed."[9] For Haraway, the cyborg offered a possible alternative to the colonized body of western culture and its ideology that denigrates women, stigmatizes indigenous people and people of color, and seeks standardization and homogenization.[10] The cyborg suggested to Haraway transcendence into hybridity and difference. For others, the cyborg suggested various forms of liberation, from the state (libertarians), from gender divisions (feminists), from labor (utopians), and from the body.[11] The cyborg shattered old categories, eliminating the dualities of mind/body, nature/culture, and male/female. As Klaus Benesch perceptively suggested, the figure of the cyborg functioned "as a foil onto which we project both the desire to improve our biological condition, that is to become more machine-like, and, at the same time, the anxiety about machines replacing the human body altogether." Because the cyborg potentially represents both perfecting human beings and making them redundant, it is an ambiguous figure that is culturally necessary in order "to negotiate the increasing technologizing of the modern world."[12] Its form remains obscure, and it might be anything from a ruthless killing machine to a genderless musical virtuoso. In some form or fashion, it might seem sublime.

But would a cyborg have any use for the sublime? Or will it prove to have been a temporary, human emotion? Conceivably, the sublime became prominent in the eighteenth century, persisted until now, but will soon fade away. A cyborg might prefer clarity, logic, and absolutes to such hallmarks of the sublime as darkness, the shapeless, the boundless, the disordered, and the tumultuous. It might have little tolerance for encounters with places that are roaring, disorded, or obscure. Would a cyborg's mental abilities include access to all nine human senses, including bodily awareness? Conceivably, sublime emotions are only possible for a consciousness inside a sensuous physical body. A self that relies on electrical sensors might not be able to experience the initial amazement of Burke's or Kant's sublime. The sublime demands not only that one sees a majestic cataract or an erupting volcano but also hears them, smells them, measures them against one's body, senses the environment they create, and synthesizes these many impressions. And even if a cyborg could experience synesthesia and be enthralled by obscurity, would it be overwhelmed or moved to philosophical reflections?

Furthermore, an essential element of the human sublime, the feeling of horror, might not be possible for a consciousness backed up in a hard drive, and essentially immortal. The natural, technological, disastrous, and martial sublimes all arise from a synthesis that may be exclusively human. Could a cyborg fully experience the dynamism of a hurricane, the acceleration of high-speed travel, the catastrophic sublime of an earthquake, or the fear of death during battle? A cyborg might understand mortality only as an idea, and it might be designed to be impervious to fear. A cyborg soldier presumably would be oblivious to pain and loud noises and immune from post-traumatic shock. Its design might preclude any experience of the martial sublime.

Yet, a cyborg might be better suited than human beings to experience forms of the intangible sublime, such as those seen through a telescope or microscope, and it might not need photographic representations to understand data beamed down from the Hubble telescope. It also might be designed to be comfortable on other planets. Compared to human beings, cyborgs might have a more direct encounter with extraterrestrial landscapes, and they might have the capacity to rapidly acquire alien languages, should they encounter intelligent life elsewhere in the universe. In short, cyborgs might take less interest in the tangible sublimes but have intense encounters with phenomena intangible to human beings.

Putting to one side speculations about cyborgs, for three centuries human beings have developed new formations of the sublime, and this trend appears likely to continue. Control of drones and integration of their flight with human vision will likely improve. Virtual reality will make it possible to sense the world from the perspective of other species, and it might evolve into a powerful replacement for cinema. Augmented reality (AR), which uses specialized glasses to add layers of information to the physical world, will also be upgraded.[13] In 2016 Microsoft developed the HoloLens, a "fully untethered, see-through, holographic computer" that enabled the creation of high-definition holograms that were "seamlessly integrating with your physical places, spaces, and things." A reviewer declared the HoloLens "a self-contained wearable Windows 10 computer. It's able to map the wearer's environment while displaying virtual objects anchored to that environment. The result is the best mixed-reality experience to date by a stand-alone, untethered device."[14] The Viennese philosopher Mark Coeckelbergh suggests that AR continues the Romantic project to re-enchant the world. It "aims to transform the world rather than escape it."[15] But will AR turn out to be disenchantment and the next sublime development emerge instead in the Metaverse?

Will philosophers take an interest in all of the sublimes, or will they continue to focus mostly on the natural sublime? Will biological research explain how the human senses interact during a sublime experience? Might there be an identifiable, physiological difference that underlies the contrasting experiences of sublime landscapes and sublime spectacles? Will VR, holography, or drones become important parts of disaster tourism? Will the increasing automation of warfare transform (or possibly undermine) the martial sublime? This seems to be occurring with drone warfare. Will experiences of the intangible sublime become more popular as they become more accessible? This seems to be the case as the public follows the Mars rovers. Will the symbiotic awareness of the environmental sublime be as widely embraced as the natural sublime? Might VR help create a breakthrough in human communication with other species? Will new technologies that mitigate global warming be considered sublime? Will the sublime be used more to enhance nationalism or internationalism? By exploring such questions, humanity will continue to stretch its imagination. It seems unlikely that we will remain limited to seven sublimes.

NOTES

PREFACE

1. Dubois, *Darkwater*, 10, 366–367.

2. Pennington, "Here to Help, Amp Up Your Awe," A3.

3. Keltner et al. "Interpersonal Relations and Group Processes: Awe and Humility," 258–269. See also Keltner and Haidt, "Approaching Awe, a Moral, Spiritual, and Aesthetic Emotion," 297–314.

4. Pennington, "Here to Help, Amp Up Your Awe," A3.

ACKNOWLEDGMENTS

1. Gross, *The Scientific Sublime*; Nye, "What Comes after the Technological Sublime?"

CHAPTER 1

1. Dickens, *American Notes*, 199–200.

2. Hepburn, "Landscape and the Metaphysical Imagination," 201.

3. Dickens, *Pictures from Italy*, 250–252.

4. Pavlovskis, *Man in an Artificial Landscape*, 1, 20, 33.

5. See Taylor, "The Awful Sublimity of the Victorian City," 436.

6. Burke, *A Philosophical Enquiry*, 53.

7. Shapshay, "Contemporary Environmental Aesthetics and the Neglect of the Sublime," 187.

8. Shusterman, "Somaesthetics and Burke's Sublime," 326.

9. Brady, *The Sublime in Modern Philosophy*, 87. See also Doran, *The Theory of the Sublime from Longinus to Kant*.

10. Henshaw, *A Tour of the Senses*, 6–7.

11. Solnit, *A Field Guide to Getting Lost*, 147.

12. Kant, *Critique of the Power of Judgment*.

13. Solnit, *A Field Guide to Getting Lost*, 151–152.

14. Kant, in Watson, *The Philosophy of Kant Explained*, 490–491.

15. Watson, *The Philosophy of Kant Explained*, 498–499. For discussion, see Brady, *The Sublime in Modern Philosophy*, 47–89.

16. Goldthwait, "Translator's Introduction," 37.

17. Nye, *American Technological Sublime*, 96–108, 241–250.

18. Cited in Kessler, *Picturing the Cosmos*, 48.

19. More than two thousand citations listed at https://scholar.google.com/citations ?view_op=view_citation&hl=en&user=W6dQ3pcAAAAJ&citation_for_view=W6d Q3pcAAAAJ:u5HHmVD_uO8C.

20. Brady, *The Sublime in Modern Philosophy*; Gray, *The Warriors*.

21. Lopez, *Arctic Dreams*, 5.

22. Devlin, "Black Hole Picture Captured for the First Time in Space Breakthrough."

23. Burke, *A Philosophical Enquiry*, 129.

24. Cited in Monk, *The Sublime*, 87.

25. Kant, cited in Licht, "Warring Opinions," 54; Burke, *A Philosophical Enquiry*, 75–79.

26. Kant, *Universal Natural History*.

CHAPTER 2

1. Andersen, "Jernbanen," 22.

2. The term "technological sublime" became widely known in the United States after two works appeared in 1965: Miller's *The Life of the Mind in America*, 295–306; and Marx's *The Machine in the Garden*, 195–207, 230–231. Both used "technological sublime" in discussing railroads and the industrial revolution, and both noted the criticism of technological nationalism by prominent writers such as Nathaniel Hawthorne and Herman Melville. Other scholars took up the term, notably Kasson, *Civilizing the Machine*, 162–172; Marchand, *Advertising the American Dream*, 280; and Sears, *Sacred Places*, 182, 192, 201. I used the term in *Electrifying America* (3, 44, 46, 58–60, 76, 84, 380, 390–391) to discuss the public response to early electric lighting and then applied it to a wide range of technologies in *American Technological Sublime*, 3, 44, 46, 58–60, 76, 84, 380, 390–391. Wilson noted in "The Postmodern Sublime": "The conversion scene of the postmodern American poet—overcoming known boundaries and humbled to belief in some saving US telos—is likely to occur not in

a natural landscape but, rather, 'under the Pyramid' of the TransAmerica Building in San Francisco or drifting within the cybernetic barrages and high-finance transactions of Wall Street," 518. See also Wilson, *American Sublime*; and Tabbi, *Postmodern Sublime*.

3. Oettermann, *The Panorama*, 101–103.

4. Oettermann, 101–112.

5. Cited in Oettermann, 124.

6. Oettermann, 145.

7. "Biblical Scenes," *Daily Ohio Statesman*, November 29, 1864; "Catherwood's Panoramas," *The Commercial Advertiser*, September 25, 1838. The surviving Mississippi panorama at the St. Louis Art Museum can be viewed at https://www.slam.org/collection/objects/841/.

8. Kaplan, *Aerial Aftermaths*, 73.

9. Thébaud-Sorger, review of *The Sublime Invention*, 488.

10. Crouch, *Lighter Than Air*, 22.

11. *New Orleans Chronicle*, July 20, 1819; reprinted, *City of Washington Gazette*, August 26, 1819.

12. Kaplan, *Aerial Aftermaths*, 69.

13. Kaplan, 76–78.

14. *Fortune Magazine* published many such images in the 1930s. See also Nye, *American Technological Sublime*, 94–95, 138–140.

15. Landau and Condit, *The Rise of the New York Skyscraper*, x. See also Nye, *American Technological Sublime*, 77–108.

16. Morshed, *Impossible Heights*.

17. Morshed, 166.

18. Deriu, "'Don't Look Down!,'" 1038.

19. Tandt, *The Urban Sublime in American Literary Naturalism*, 245–247.

20. The new skyscrapers in Dubai, Taiwan, Kuala Lumpur, and Shanghai are "beautiful and terrifying at the same time. In part, this is the sublime terror of the abyss and the peak, of elevation and descent. It is also the sense of being diminished by something on an entirely inhuman scale, with the possibility that a small squashy body is reduced to insect-like insignificance in the face of such mass." Parker, "Vertical Capitalism," 218–219.

21. See Nye, *American Illuminations*, 1–34.

22. Nye, 109–132.

23. Nye, *American Technological Sublime*, 296.

24. Nye, *When the Lights Went Out*, 10–13, 19, 58, passim.

25. Marlowe, "Natural and Technological Wonders," 24–25.

26. Rabinovitz, *Electric Dreamland*, 133.

27. Jackson, "The Abstract World of the Hot-Rodder," 146–147.

28. McClelland, *Building the National Parks*, 176–177.

29. McClelland, 188–189.

30. Sutter, *Driven Wild*, xi, 120–121.

31. Mauch and Zeller, eds., *The World Beyond the Windshield*, 92.

32. Mauch and Zeller, 86, 92.

33. Cited in Mauch and Zeller, 92.

34. Based on discussions with Thomas Zeller on his research dealing with scenic highways.

35. Krapp, "Nomads of the Technical Sublime," 205–219.

36. Missal, *Seaway to the Future*.

37. "Industrial Sublime: Modernism and the Transformation of New York's Rivers 1900–1940," Hudson River Museum, October 4, 2013. https://www.youtube.com/watch?v=vKoEEHKWde0&ab_channel=HudsonRiverMuseum.

38. Gold, "Same New York Rivers, but All Else Was New."

39. Nye, *America's Assembly Line*, 46–48.

40. Nye, 44–45.

41. Marx, *The Machine in the Garden*, 355–356.

42. Littmann, "The Production of Goodwill," 79.

43. Littmann, 81.

44. LeCain, *Mass Destruction*, 109–110.

45. Leech, "Protest, Power, and the Pit," 28. See also, "Berkeley Open Pit Draws Thousands of Tourists," 1; "Berkeley Pit One of Montana's Most Spectacular Sights," 4.

46. Quotation translated and shared by Victor Seow via email, November 1, 2017.

47. LeCain, *Mass Destruction*, 204.

48. Nye, *American Technological Sublime*, 123–128; Dreiser, *A Hoosier Holiday*, 17–18.

49. Stilgoe, *Metropolitan Corridor*, 79–81.

50. Aurand, *The Spectator and the Topographical City*, 118.

51. Younger, *Industry in Art*.

52. Nye, *American Technological Sublime*, 125–127.

53. Cited in Price, *Coal Cultures*, 159.

54. Edward Burtynsky, "Exploring the Residual Landscape."

55. Baichwal, *Manufactured Landscapes*.

56. Brady, *The Sublime in Modern Philosophy*, 173. On whether skyscrapers or other architecture is sublime, see Brady, 142–144.

57. Bell and Lyall, *The Accelerated Sublime*, 110–111.

58. Orvell, *Photography in America*, 30.

59. Pankhuri, "Train Crushes 2 Boys Clicking Selfies in Delhi."

60. Chlu, "More than 250 People Die Worldwide Taking Selfies."

61. "Selfie Deaths," BBC News.

62. Du Preez, "Sublime Selfies," 746, 748.

63. Du Preez, 756.

64. Howlett, "Americans Driving to Distraction as Multitasking on Road Rises." See also Galant, "Driven to Distraction."

65. Nye, *American Technological Sublime*, 291–296.

CHAPTER 3

1. *The Great Chicago Fire of 1871*, 38–40.

2. See Hoffman and Oliver-Smith, *Culture and Catastrophe*; UNISDR, *Global Assessment Report*; Finn, *Documenting Aftermath*.

3. Gray, *The Warriors*, 29.

4. "Dreadful Fire," *Gazette of the United States*, February 7, 1803, 3.

5. "Remarkable Conflagration," *The Portsmouth Journal of Literature and Politics*, July 11, 1840, 1.

6. Burke, *A Philosophical Enquiry*, 43.

7. Burke, 43.

8. McCullough, *The Johnstown Flood*, 224–225.

9. Smith, Ottoni-Wilhelm, and Scharf, "The Donation Response to Natural Disasters," 2.

10. Smith, Ottoni-Wilhelm, and Scharf, 93–95, 113–114, 258.

11. Oettermann, *The Panorama*, 208–209.

12. "William Gallagher," *Anthology of Fire Narratives*, accessed October 6, 2021, https://www.greatchicagofire.org/anthology-of-fire-narratives/william-gallagher.

13. Cited in Rozario, *The Culture of Calamity*, 102.

14. Rozario, 101–108.

15. Cited in Rozario, 113.

16. Genthe, "Sacramento Street, San Francisco, 1906."

17. Klauber, "Two Days in San Francisco, 1906."

18. Rozario, *The Culture of Calamity*, 103.

19. *The Great Chicago Fire of 1871*, 51.

20. Smith, "Faith and Doubt," 145.

21. Smith, 129–133.

22. McCullough, *The Johnstown Flood*, 217–218.

23. Yablon, *Untimely Ruins*, 193.

24. Yablon, 198, 205–206.

25. Steinberg, *Acts of God*, 4.

26. Berry, *The Complete Pompeii*, 25–27.

27. Cited in Kovacs, "Pompeii and Its Material Reproductions," 25–49.

28. Dickens, *Pictures from Italy*, 245.

29. Rabinovitz, *Electric Dreamland*, 54–55.

30. Manback, *Brooklyn*, n.p.

31. Valance, "Destructive Re-creations," 121–128.

32. Rabinovitz, *Electric Dreamland*, 54.

33. Cited in Vardi, "Auto Thrill Shows and Destruction Derbies, 1922–1965," 22.

34. Rabinovitz, *Electric Dreamland*, 8.

35. Pezzullo, "'This is the only tour that sells,'" 100.

36. Peebles, "Toxic Sublime," 373–392.

37. Peebles, 375.

38. Baudrillard, *America*, 17.

39. DeLillo, *White Noise*, 127.

40. DeLillo, 127.

41. Langford, "Seeing Only Corpses," 47.

42. Orvell, "Photographing Disaster," 647.

43. Orvell, 660.

44. Wood, "My Atomic Holiday."

45. O'Connell, "Journey to the End of the World," 36, 40; Yankovska and Hannam, "Dark and Toxic Tourism in the Chernobyl Exclusion Zone," 929, passim.

46. Goatcher and Brunsden, "Chernobyl and the Sublime Tourist," 115–137.

47. White and Frew, "Dark Tourism and Place Identity."

48. Barringer, "Despite Cleanup at Mine, Dust and Fear Linger"; Nye, "Superfund Sites as Anti-landscapes," 283–305.

49. Nye and Elkind, *Antilandscapes*.

50. McCarthy, *The Road*.

51. Orvell, *Empire of Ruins*, 207–216.

52. Nurmis, "Visual Climate Change Art, 2005–2015," 501–516.

53. Dunaway, "Seeing Global Warming," 9–31.

54. Cited in Nurmis, "Between Dread and Delight."

55. Smith, "Marc Quinn's 'The Toxic Sublime.'"

56. Nixon, *Slow Violence*, 66.

57. Nye, *American Technological Sublime*, 234–237.

58. Wakefield, "Infrastructures of Liberal Life," 4.

CHAPTER 4

1. Wohl, *A Passion for Wings*, 69–96; D. H. Lawrence, September 9, 1915, in Zytaruk and Boulton, *The Letters of D. H. Lawrence*, vol. 2, 389–390.

2. Gray, *The Warriors*, 32.

3. Cited in Crouch, *Lighter Than Air*, 54.

4. Crouch, 59.

5. One might argue that dangerous recreation activities such as deep-sea diving, hang gliding, and bungee jumping could be considered a subcategory of the martial sublime. I leave this question for others to pursue.

6. Burke, *A Philosophical Enquiry*, 75–76, 79. Burke also argues that in *The Iliad* the strength of the Greeks (notably Achilles) partakes of the sublime, while the Trojans have amiable virtues that dispose a reader to love and pity them, 143–144.

7. See Sledge, *With the Old Breed at Peleliu and Okinawa*, 63.

8. Sledge, 66. See also 75, 211.

9. Orvell, *American Photographs*, 64–65.

10. Adas, *Machines as the Measure of Man*, 365.

11. Orvell, "Cultural Studies, Visual Studies, Historical Studies," 100.

12. Lilienthal, *Sacred Ground*.

13. Reprinted as "The French Army," *The Pittsfield Sun*, June 16, 1859.

14. Dawes, *The Language of War*, 34.

15. Holman, "The Meaning of Hendon," 139.

16. Jacobs, "On the Frontline of War-Zone Tourism"; Kamin, "The Rise of Dark Tourism."

17. Higgins, "No Bed, No Breakfast, but 4-Star Gunfire."

18. "A Reprinted Letter," *The Connecticut Courant*, November 26, 1816.

19. Gray, *The Warriors*, 32.

20. Dawes, *The Language of War*, 74.

21. Gray, *The Warriors*, 33.

22. Gray, 36.

23. Gray, 36–37.

24. Licht, "Warring Opinions," 54.

25. Keegan, *The Face of Battle*, 203–231.

26. Ferguson, "The Sublime and the Subliminal: Modern Identities and the Aesthetics of Combat," 14.

27. "Excerpt of a letter from an officer in the American army to his friend in Farmington," *American Mercury*, September 6, 1814.

28. Cited in Licht, "Warring Opinions," 24–25.

29. Licht, 54. Kant also thought that a long peace tended "to make prevalent a mere[ly] commercial spirit, and along with it base selfishness, cowardice, and softness, and to debase the way of thinking of that people."

30. Junger, *War*, 53–66, passim.

31. Burke, *A Philosophical Enquiry*, 36.

32. Burke, 75.

33. Andrews, *History of the Campaign of Mobile*, 151.

34. Ferguson, "The Sublime and the Subliminal," 15.

35. Romains, *Verdun*, 93.

36. Burke, *A Philosophical Enquiry*, 127.

37. Burke, 127.

38. Junger, *War*, 34–35.

39. Ripley, *The Unthinkable*, 58.

40. Licht, "Warring Opinions," 18.

41. Licht, 6–7, 8–9.

42. Caputo, *A Rumor of War*, xvi–xvii.

43. O'Brien, "How to Tell a True War Story," 87. For examples, see Bjerre, "'Let Me See Your War Face,'" 64–72.

44. Licht, "Warring Opinions," 173.

45. "Ghastly Planes Seen by Fokker," *Trenton Evening Times*, August 25, 1922. On Douhet, see Sherry, *The Rise of American Air Power*, 22–27.

46. Licht, "Warring Opinions," 8.

47. Nye, *When the Lights Went Out*, 48–52.

48. Allison, *Destructive Sublime*.

49. Friedrich and Brown, *The Fire*.

50. Colem Hemez, "Photographic Visions of the Atomic Sublime," 17.

51. Hales, "The Atomic Sublime," 10.

52. Cited in Hales, 12.

53. Hales, 16.

54. Winkler, *Life Under a Cloud*, 92.

55. C. Gallagher, *American Ground Zero*, 217.

56. Wood, "My Atomic Holiday."

57. Wray, "A Blast from the Past," 472.

58. Gowans, "WSMR [White Sands Missile Range]," 2, 5.

59. Gowans, 9.

60. Quoted in Terry, "Stockhausen, Karlheinz," 1. See also Spinola, "Monstrous Art;" Schechner, "9/11 as Avant-Garde Art?," 1820–1829.

61. Orvell, "After 9/11," 7.

62. Debrix, "The Sublime Spectatorship of War," 770–771.

63. Ferguson, "The Sublime and the Subliminal," 12.

64. Hemingway, "Soldier's Home," 94–95.

65. Ellison and Monaugh, "Simon Norfolk in Interview," 203.

66. O'Brien, "How to Tell a True War Story," 87.

CHAPTER 5

1. Clewis, "What's the Big Idea? On Emily Brady's Sublime," 111.

2. Clewis, 113.

3. Heath, "Longinus and the Ancient Sublime," 12.

4. Golinski, "Joseph Priestley and the Chemical Sublime in British Public Science," 122.

5. Gross, *The Scientific Sublime*.

6. Cited in Gross, 16–17.

7. Johnston, *Holograms*, 20, 31, 42; Johnston, *Holographic Visions*, 393.

8. Johnston, *Holographic Visions*, 189.

9. Johnston, 313, 317.

10. Cited in Johnston, 319.

11. Johnston, 408.

12. Brady, *The Sublime in Modern Philosophy*, 128–129.

13. Nye, *American Technological Sublime*, 291.

14. Boczkowska, "Spaceflight as the (Trans)National Spectacle."

15. Wasson, "The Networked Screen," 88.

16. Wormbs, "Sublime Satellite Imagery as Environing Technology," 77–92.

17. Cited in Gray, "Drones, War, and Technological Seduction," 954.

18. Chayka, "The Troubling Contradictions of Dronestagrams."

19. Vanberburg, "Drone Art," 8.

20. For discussion, see Nye, *Electrifying America*.

21. "About the Hubble Space Telescope," last updated December 18, 2018, https://www.nasa.gov/mission_pages/hubble/story/index.html.

22. Zimmerman, *The Universe in a Mirror*, 178.

23. Kessler, *Picturing the Cosmos*, 55.

24. Kessler, 8.

25. Cited in Kessler, 62.

26. Kessler, 63.

27. Kessler, 147–148.

28. Kessler, 192.

29. Mars Exploration Program, accessed October 6, 2021, https://mars.nasa.gov /#mars_exploration_program/3.

30. Greshko, "The Mars Rover Opportunity Is Dead."

31. Clancey, "Becoming a Rover," 114.

32. Vertesi, "Seeing Like a Rover," 397.

33. Clancey, "Becoming a Rover," 116.

34. Bresina et al., "Activity Planning for the Mars Exploration Rovers."

35. Clancey, "Becoming a Rover," 121.

36. Vertesi, "Seeing Like a Rover," 396.

37. Clancey, "Becoming a Rover," 118.

38. Helmreich, "Intimate Sensing," 129, 138–147.

39. Natarajan, "At Long Last, a Glimpse of a Black Hole."

40. Overbye, "Darkness Visible, Finally."

41. Fletcher, "How Do You Take a Picture of a Black Hole?"

42. Gross, *The Scientific Sublime*, 25–62.

CHAPTER 6

1. See Nye, *American Technological Sublime*, 61; Howley, *Drones*, 8–9.

2. On utopian visions of the Internet, see Mosco, *The Digital Sublime*, 106, passim; Stone, *The War of Desire and Technology*; Reingold, *The Virtual Community*, 111, 130; Negroponte, *Being Digital*, 238; Coeckelbergh, *New Romantic Cyborgs*, 144–147.

3. On dystopian views of the Internet, see Roszak, *The Cult of Information*, 202–233; Zuboff, *The Age of Surveillance Capitalism*.

4. Burke, *A Philosophical Enquiry*, 66.

5. Goldthwait, "Translator's Introduction," 37.

6. de Mul, "The (Bio)Technological Sublime," 32–40.

7. Willim, "Imperfect Imaginaries," 65–66, 70; Nye, *Image Worlds*, 55, passim; Nye, *American Technological Sublime*, 114–115.

8. Littmann, "The Production of Goodwill," 71–84.

9. Nye, *American Technological Sublime*, 101.

10. Rosenfield, "We've Reached the Summit."

11. Similar strategies appeared in a video Google later posted online: https://www .google.com/about/datacenters/inside/streetview/.

12. Thanks to Tom Misa for sharing his knowledge of early simulations and virtual reality.

13. Belisle, "Immersion," 254; Allen, "A Brief History of Immersion."

14. Orvell, "Virtual Culture and the Logic of American Technology," 12–27. See also Belisle, "Immersion," 250.

15. Belisle, 255.

16. Greengard. *Virtual Reality*, 18.

17. Frana, interview with Carl Machover, 20.

18. Lanier, *You Are Not a Gadget*, 185.

19. Fisher et al., "Virtual Interface Environment Workstations."

20. Frana, interview with Carl Machover, 21.

21. Campbell-Kelly et al., *Computer*, 143–151.

22. Ellis, "Nature and Origins of Virtual Environments," 321–322, 326–327.

23. Ellis, 324–325. See also Greengard, *Virtual Reality*, 215.

24. Ellis, "Nature and Origins of Virtual Environments," 328.

25. Friedman, *Electric Dreams*, 167.

26. Gibson, *Burning Chrome*, 196–197.

27. Gibson, *Neuromancer*, 35.

28. Nye *American Technological Sublime*, 107.

29. Leiner et al. "Brief History of the Internet."

30. Weiskel, *The Romantic Sublime*, discussed in Dorsen, "The Sublime and the Digital Landscape," 55–67.

31. Dorsen, 60–61.

32. King and Borland, *Dungeons and Dreamers*, 151.

33. Ryan, "Narrative," 336.

34. King and Borland, *Dungeons and Dreamers*, 152.

35. King and Borland, 162–170.

36. King and Borland, 211, 220.

37. King and Borland, 223.

38. Shinkle, "Video Games and the Digital Sublime," 95.

39. Friedman, *Electric Dreams*, 121.

40. Baron, "Pressures on Play," 77.

41. Baron, 74, 80, 115.

42. Krapp, "Control," 78.

43. Friedman, *Electric Dreams*, 136.

44. Friedman, 138, 139.

45. Myers, "Simulations," 396.

46. Kirshenbaum, "Krigsspiel," 280.

47. See "Interview with James F. Dunnigan," in Huntemann and Payne, *Joystick Soldiers*, 67–70; Myers, "Simulations," 396; Allen, *America's Digital Army*, 120.

48. Cited in Shaw, *Predator Empire*, 73.

49. Miller and Thorpe, "SIMNET," 1115.

50. Miller and Thorpe, 1114.

51. SIMNET 2 was used in complex materials handling systems, using automatic guided vehicles; see McHaney, "Development of a Generic Simnet II Simulation Package."

52. Miller and Thorpe, "SIMNET," 1121–1122.

53. Allen, *America's Digital Army*, 121.

54. Miller and Thorpe, "SIMNET," 1121.

55. Gresham, "Gulf War."

56. Allen, *America's Digital Army*, 5. See also Nichols, "Target Acquired."

57. Nichols, 42.

58. Howley, *Drones*, 177–178.

59. Brant, *Grounded*; Niccol, *Good Kill*.

60. Cited in Howley, *Drones*, 184.

61. Sluka, "Drones in the Tribal Zone," 21–33.

62. Yost, interview with Donn Parker, 20.

63. Dorsen, "The Sublime and the Digital Landscape," 67.

64. Zuboff, *The Age of Surveillance Capitalism*, 510.

65. Ruiz, "Darfur Is Dying," 2006.

66. Peña et al., "Immersive Journalism," 299.

67. Slater et al., "Inducing Illusory Ownership of a Virtual Body," 214–220.

68. Lanier, *The Dawn of the New Everything*, 55.

69. Williams, *Red*, 4.

70. Lanier, *The Dawn of the New Everything*, 140–141.

71. Johnson, "Natural Magic," 52–53.

72. Lanier, *You Are Not a Gadget*, 9.

73. Lanier, 11–12.

74. Strand, *Inventing Niagara*, 302.

75. See Vanian, "Amazon Takes a Trip in Augmented and Virtual Reality;" MacRumors Staff, "Apple Glasses;" Discover, "GoogleVR,"

76. Greengard, *Virtual Reality*, 119–137.

77. Burke, *A Philosophical Enquiry*, part 2, section 1.

78. Turkle, *Simulation and Its Discontents*, 73.

79. Turkle, 74–75.

80. Yu et al. "Skin-Integrated Wireless Haptic Interfaces," 473–475.

81. Lanier, *Dawn of the New Everything*, 129–139.

82. Greengard, *Virtual Reality*, 20–23, passim.

83. Kemeny et al., "New VR Navigation Techniques to Reduce Cybersickness," 48–53.

84. Lopez, *Horizon*, 24–25.

85. Burke, *A Philosophical Enquiry*, part 2, section 4.

86. Gordon, "Can Virtual Nature Be a Good Substitute for the Great Outdoors? The Science Says Yes."

CHAPTER 7

1. McKibben, "Designer Genes," 266.

2. Hitt, "Toward an Ecological Sublime," 606–607.

3. Hitt, 609, 617.

4. Nye, *Conflicted American Landscapes*, 135, 152–154, 160–163.

5. Cialdella, "Looking for Nature in the Rust Belt," 111.

6. Gross, *The Scientific Sublime*, 254–255.

7. Oppermann and Iovino, "The Environmental Humanities and the Challenges of the Anthropocene."

8. Bolster, *The Mortal Sea*, 45–46, 68, 109, 203, 216, 285–290.

9. Lopez, *Arctic Dreams*, 11, 25.

10. Lopez, 39.

11. Lopez, 200.

12. Davy cited in Heingman, "The Style of Natural Catastrophes," 104–105.

13. Abrams, *Natural Supernaturalism*, 59–61; Hartman, *Wordsworth's Poetry*, 240–257.

14. Emerson, "Country Life," 61.

15. Brady, *The Sublime in Modern Philosophy*, 190–191.

16. Brady, 191.

17. Hepburn, "Landscape and the Metaphysical Imagination," 192.

18. Brady, *The Sublime in Modern Philosophy*, 192; emphasis in original.

19. Brady, 206.

20. Hata et al., "Geographic Variation in the Damselfish-Red Alga," 185.

21. Mynott, "Newly Discovered Hermit Crab Species Lives in 'Walking Corals.'"

22. Pringle et al., "Spatial Pattern Enhances Ecosystem Functioning in an African Savanna," e1000377.

23. Malmstrom, "Ecologists Study the Interactions of Organisms and Their Environment," 88.

24. Peterson, *Jane Goodall*.

25. Horner, "The Cultural Mind of Chimpanzees," 101–115.

26. Linzey, ed., *The Global Guide to Animal Protection*, 43–44.

27. Wohlleben, *The Hidden Life of Trees*, 20.

28. Kimmerer, *Braiding Sweetgrass*, 20.

29. Kimmerer, 18–21.

30. Suzuki and McConnell, *The Sacred Balance*, 142.

31. Ackerman, *The Human Age*, 153.

32. Economides, *The Ecology of Wonder in Romantic and Postmodern Literature*, 30, 131–133.

33. Whittle, "Ambitious New Plan to Save Atlantic Salmon Has a Big Price Tag." See also United States Fish & Wildlife Service, "Migratory Fish Restoration Benefits Everyone."

34. Conniff, "Rebuilding the Natural World."

35. Jørgensen, *Recovering Lost Species in the Modern Age*, 67–76, 84–88, 135.

36. Overgaard, "To Keep African Swine Fever Out, Denmark Is Planning a Southern Boar(der) Fence."

37. Kolbert, *The Sixth Extinction*; Jørgensen, "Rethinking Rewilding."

38. Bonneuil and Fressoz, *The Shock of the Anthropocene*, 85.

39. Jørgensen, *Recovering Lost Species in the Modern Age*, 113–117; Ackerman, *The Human Age*, 151–152. Zimmer, "A New Company with a Wild Mission."

40. Nixon, *Slow Violence*, 64–65.

41. Emmett and Nye, *The Environmental Humanities*. See also Wilson, "The Sublime Anthropocene," 155–174.

42. Worster, *Nature's Economy*.

43. Bonneuil and Fressoz, *The Shock of the Anthropocene*, 170–176, passim.

44. Miller, *An Environmental History of Latin America*, 15–18.

45. Sonter et al., "Mining Drives Extensive Deforestation in the Brazilian Amazon," 1013.

46. Scranton, *Learning to Die in the Anthropocene*, 19.

47. Dunaway, *Seeing Green*, 12.

48. Dunaway, 38, 145, 258–267.

49. Emmett and Nye, *The Environmental Humanities*.

50. Beck, *Risk Society*, 81.

51. Cronon, "The Trouble with Wilderness," 69–90. See also Evernden, *The Social Creation of Nature.*

52. Owen, *Green Metropolis*, 43–48, 177–190, passim.

53. Larsen, "The Living Archive and the Sublime Nature of the Anthropocene," 223–238.

54. Rozelle, *Ecosublime*, 60.

55. Morton, *Being Ecological*, 215.

56. Morton, *Ecology without Nature*, 46.

57. Woo, "Ecomimesis," 255–266.

58. Woo, 261.

59. See Economides, *The Ecology of Wonder in Romantic and Postmodern Literature*, 136–137.

60. Kimmerer, *Braiding Sweetgrass*, 368.

61. Kimmerer, 140.

CHAPTER 8

1. Heath, "Longinus and the Ancient Sublime," 13.

2. See Frierson, "Introduction," xxix–xxx.

3. Renan, "What Is a Nation?"; Anderson, *Imagined Communities*; Hobsbawm and Ranger, *The Invention of Tradition.*

4. Volcanoes of Kamchatka, https://whc.unesco.org/en/list/765/.

5. Steller, *Steller's History of Kamchatka.*

6. Sears, *Sacred Places*; McKinsey, *Niagara Falls;* Haw, *The Brooklyn Bridge.*

7. Sayre, "If Jefferson Had Visited Niagara Falls," 141–162.

8. Nye, *American Technological Sublime*, 9–11, 100–106, 241–250.

9. Haselstein, "Seen from a Distance," 405–421.

10. Lyotard, "The Sublime and the Avantgarde," 95.

11. Nye, *American Technological Sublime*, 24–25, 40–43, 75–76, 155–160, passim.

12. Brown, "The First American Sublime," 166.

13. Durkheim, "The Elementary Forms of Religious Life," 153.

14. For discussion of such contradictions, see Nye, *Conflicted American Landscapes.*

15. Duffy, *The Landscapes of the Sublime, 1700–1830.*

16. Jager, "Picturing Nations," 123–124.

17. Bell and Lyall, *The Accelerated Sublime*, 34.

18. Dobraszczyk, "Sewers, Wood Engraving and the Sublime," 351.

19. Fyfe, "Illustrating the Accident," 61–64.

20. Agar, "Technology, Environment and Modern Britain," 14.

21. In the United States, the term "technological sublime" became more widely known after publication of Miller's *The Life of the Mind in America*, 295–306; and Marx's *The Machine in the Garden*, 195–207, 230–231.

22. Styran and Taylor, *This Colossal Project.*

23. Willis, "William Notman."

24. Hall et al., *The World of William Notman.*

25. Rodgers, "Constructing Beauty," 72–91. Another example is the Quebec Bridge over the Saint Lawrence River, completed in 1919, after a disastrous collapse during construction in 1907.

26. Stein, "All That Is Solid Melts into Air Travel."

27. Stein, 77.

28. Nayar, "The Imperial Sublime," 58.

29. Nayar, 80.

30. Chatterjee, *The Great Indian Railways*, ch. 1, n. 18.

31. Kupensky, "The Soviet Industrial Sublime," 8.

32. Kupensky, 10–12.

33. Buck-Morss, *Dreamworld and Catastrophe*, 185. Discussed in Kupensky, "The Soviet Industrial Sublime," 12–13.

34. Lifton, *Witness to an Extreme Century*, 307–317.

35. Brady, *The Sublime in Modern Philosophy*, 204.

36. However, many Chinese fear the "Internet addiction" of their children, which they see as "a formidable threat to the moral education, family life, schooling, career and physical well-being of youth." Tarantino, "Toward a New 'Electrical World,'" 194, 197, passim.

37. Costa, "Estado, tecnologia y sociedad en las infraestructuras."

38. "South America's Drought-Hit Paraná River at 77-Year Low," BBC News.

39. Hecht, *The Radiance of France.*

CHAPTER 9

1. Brady, *The Sublime in Modern Philosophy*, 58–62, passim.

2. Gilkey, "The Spiritual Uplift of Scenery as Exemplified by the Grand Canyon," 138.

3. Missal, *Seaway to the Future*, 188–193.

4. Munk et al., "Lithium Brines," passim.

5. Gray, *The Warriors*, 23.

6. Ackerman, *The Human Age*, 172.

7. Macfarlane, *Underland*, 15–16.

8. Garrard, "Memories of Snow."

9. On the collapse of the distinction between nature and culture, see Emmett and Nye, *The Environmental Humanities*, 9–11, 39–42, 95–99, 145–149.

10. Brady, *The Sublime in Modern Philosophy*, 191.

11. Leopold, "The Round River," 197.

EPILOGUE

1. Mayor, *Gods and Robots*.

2. Kang, *Sublime Dreams of Living Machines*, 55.

3. Kang, 96.

4. Kang, 185–290.

5. Clynes and Kline, "Cyborgs and Space," 27.

6. Cited in Channell, *The Vital Machine*, 133.

7. Haraway, "Morphing in the Order," 204.

8. Haraway, "Cyborgs to Companion Species," 299.

9. Haraway, "A Manifesto for Cyborgs," 11. On Haraway as romantic, see Coeckelbergh, *New Romantic Cyborgs*, 183.

10. Kroker, *Body Drift*, 113–117.

11. Dery, *Escape Velocity*, 232–233; Brooks, *Flesh and Machines*, 189.

12. Benesch, "Technology, Art, and the Cybernetic Body," 381.

13. Greengard, *Virtual Reality*, 196–199.

14. Furlan, "See the Future through Microsoft's HoloLens Augmented-Reality Glasses."

15. Coeckelbergh, *New Romantic Cyborgs*, 200.

BIBLIOGRAPHY

Abrams, M. H. *Natural Supernaturalism*. New York: W. W. Norton, 1971.

Ackerman, Diane. *The Human Age*. New York: W. W. Norton, 2014.

Adas, Michael. *Machines as the Measure of Man*. Ithaca: Cornell University Press, 1989.

Agar, Jon. "Technology, Environment and Modern Britain: Historiography and Intersections." In *Histories of Technology, the Environment and Modern Britain*, ed. Jon Agar and Jacob Ward, 1–21. London: UCL Press, 2018.

Allen, Patrick T. "A Brief History of Immersion, Centuries before VR." May 16, 2018. www.bradford.ac.uk/news/archive/2018/a-brief-history-of-immersion-centuries -before-vr.php.

Allen, Robertson. *America's Digital Army: Games at Work and War*. Lincoln: University of Nebraska Press, 2017.

Allison, Tanine. *Destructive Sublime: World War II in American Film and Media*. New Brunswick: Rutgers University Press, 2018.

Andersen, H. C. "Jernbanen." *En Digters Bazar* (1842). Reprinted in *H. C. Andersen: Romaner og Rejseskildringer*, vol. 6, ed. H. Topsøe-Jensen. Copenhagen: Gyldendal, 1944.

Anderson, Benedict. *Imagined Communities: Reflections on the Origins and Spread of Nationalism*. London: Verso, 1983.

Andrews, Brevet Major-General C. C. *History of the Campaign of Mobile*. New York: D. Van Nostrand, 1867.

Aurand, Martin. *The Spectator and the Topographical City*. Pittsburgh, PA: University of Pittsburgh Press, 2006.

Baichwal, Jennifer, dir. *Manufactured Landscapes*. Foundry Films, 2006. 90 min.

Baron, Robert John. "Pressures on Play: Rhetoric, Virtual Environments, and the Design of Experience in Virtual World Computer Games." PhD dissertation, University of Minnesota, 2012.

Barringer, Felicity. "Despite Cleanup at Mine, Dust and Fear Linger." *New York Times*, April 12, 2004.

Baudrillard, Jean. *America*. New York: Verso, 1990.

Beck, Ulrich. *Risk Society*. Los Angeles: Sage, 1992.

Belisle, Brooke. "Immersion." In *Debugging Game History: A Critical Lexicon*, ed. Henry Lowood and Raiford Guins, 247–258. Cambridge, MA: MIT Press, 2016.

Bell, Claudia, and John Lyall. *The Accelerated Sublime: Landscape, Tourism, and Identity*. Westport: Praeger, 2002.

Benesch, Klaus. "Technology, Art, and the Cybernetic Body: The Cyborg as Cultural Other in Fritz Lang's 'Metropolis' and Philip K. Dick's 'Do Androids Dream of Electric Sheep?'" *Amerikastudien / American Studies* 44, no. 3 (1999): 379–392.

"Berkeley Open Pit Draws Thousands of Tourists." *Anaconda Copper Trailsman*, August 15, 1956.

"Berkeley Pit One of Montana's Most Spectacular Sights." *Anaconda Copper Trailsman*, April 1, 1960.

Berry, Joanne. *The Complete Pompeii*. London: Thames & Hudson, 2007.

Bjerre, Thomas Ærvold. "'Let Me See Your War Face' amerikaniseringen af Danmarks nye krig i film og litteratur." *Økomomi & Politik* 90, no. 1 (April 2017): 64–72.

Boczkowska, Kornelia. "Spaceflight as the (Trans)National Spectacle: Transforming Technological Sublime and Panoramic Realism in Early IMAX Space Films." In *Multiculturalism, Multilingualism and the Self: Literature and Culture Studies*, ed. Jacek Mydla, Małgorzata Poks, Leszek Drong, 123–135. Cham, Switzerland: Springer, 2017.

Bolster, W. Jeffrey. *The Mortal Sea: Fishing the Atlantic in the Age of Sail*. Cambridge, MA: Harvard University Press, 2012.

Bonneuil, Christophe, and Jean-Baptiste Fressoz. *The Shock of the Anthropocene*. London: Verso, 2016.

Bosco, Ronald A., and Joel Myerson, eds. *The Later Lectures of Ralph Waldo Emerson, Volume 2: 1855–1891*. Athens: University of Georgia Press, 2010.

Brady, Emily. *The Sublime in Modern Philosophy*. Cambridge: Cambridge University Press, 2013.

Brant, George. *Grounded*. London: Oberon Books, 2013.

Bresina, John L., Ari K. Jónsson, Paul H. Morris, and Kanna Rajan. "Activity Planning for the Mars Exploration Rovers." Paper presented at the Fifteenth International Conference of Automated Planning and Scheduling (ICAPS 2005), 10 pp.

Brooks, Rodney A. *Flesh and Machines: How Robots Will Change Us*. New York: Vintage Books, 2003.

Brown, Chandos Michael. "The First American Sublime." In *The Sublime*, ed. Timothy M. Costelloe, 147–170. Cambridge: Cambridge University Press, 2012.

Buck-Morss, Susan. *Dreamworld and Catastrophe: The Passing of Mass Utopia in East and West*. Cambridge, MA: MIT Press, 2002.

Burke, Edmund. *A Philosophical Enquiry into the Origin of Our Ideas of the Sublime and Beautiful*. Oxford: Oxford University Press, 1990.

Burtynsky, Edward. "Exploring the Residual Landscape." Accessed October 5, 2021. https://www.edwardburtynsky.com/about/statement.

Campbell-Kelly, Martin, William Aspray, Nathan Ensmenger, and Jeffrey R. Yost. *Computer: A History of the Information Machine*. Boulder, CO: Westview Press, 2014.

Caputo, Philip. *A Rumor of War*. New York: Holt, Rinehart and Winston, 1977.

Castle, Terry. "Stockhausen, Karlheinz: The Unsettling Question of the Sublime." *New York*, August 27, 2011.

Channell, David F. *The Vital Machine: A Study of Technology and Organic Life*. New York: Oxford University Press, 1991.

Chatterjee, Arup K. *The Great Indian Railways: A Cultural Biography*. New Delhi: Bloomsbury Publishing, 2019.

Chayka, Kyle. "The Troubling Contradictions of Dronestagrams." *New Republic*, April 19, 2017. https://newrepublic.com/article/141671/troubling-contradictions-drone -photography-dronestagram-review.

Chlu, Allyson. "More Than 250 People Die Worldwide Taking Selfies." *Washington Post*, October 3, 2018.

Cialdella, Joseph Stanhope. "Looking for Nature in the Rust Belt: The Sublime of Andrew Moore's Detroit Disassembled." *Environmental History* 19 (January 2014): 110–116.

Clancey, William J. "Becoming a Rover." In *Simulation and Its Discontents*, ed. Sherry Turkle, 107–128. Cambridge, MA: MIT Press, 2009.

Clewis, Robert R. "What's the Big Idea? On Emily Brady's Sublime." *Journal of Aesthetic Education* 50, no. 2 (Summer 2016): 104–118.

Clynes, Manfred, and Nathan S. Kline. "Cyborgs and Space." *Astronautics* (September 1960): 26–27, 74–76.

Coeckelbergh, Mark. *New Romantic Cyborgs: Romanticism, Information Technology, and the End of the Machine*. Cambridge, MA: MIT Press, 2017.

Conniff, Richard. "Rebuilding the Natural World: A Shift in Ecological Restoration." *Yale Environment 360*, March 2014. e360.yale.edu/features/rebuilding_the_natural _world_a_shift_in_ecological_restoration.

Costa, Camila. "Estado, tecnologia y sociedad en las infraestructuras que atravie-sas el rio Paraná (Argentina) en la segunda mitad del siglo XX." *Registros* 14, no. 1 (2018): 141–157.

Costelloe, Timothy M., ed. *The Sublime*. Cambridge: Cambridge University Press, 2012.

Cronon, William. "The Trouble with Wilderness: Or, Getting Back to the Wrong Nature." In *Uncommon Ground: Rethinking the Human Place in Nature*, 69–90. New York: W. W. Norton, 1995.

Crouch, Tom D. *Lighter Than Air: An Illustrated History of Airships*. Baltimore: Johns Hopkins University Press, 2009.

Dawes, James. *The Language of War*. Cambridge, MA: Harvard University Press, 2002.

Debrix, François. "The Sublime Spectatorship of War: The Erasure of the Event in America's Politics of Terror and Aesthetics of Violence." *Millennium: Journal of International Studies* 34, no. 3 (2006): 767–791.

DeLillo, Don. *White Noise*. London: Picador, 1985.

de Mul, Jos. "The (Bio)Technological Sublime." *Diogenes* 59, no. 1–2 (2013): 32–40.

Deriu, Davide. "'Don't Look Down!' A Short History of Rooftopping Photography." *Journal of Architecture* 21, no. 7 (2016): 1–29.

Dery, Mark. *Escape Velocity*. New York: Grove Press, 1996.

Devlin, Hannah. "Black Hole Picture Captured for the First Time in Space Break-through." *The Guardian*, April 10, 2019.

Dickens, Charles. *American Notes*. New York: Fromm International Publishing, 1985.

Dickens, Charles. *Pictures from Italy*. London: Bradbury & Evans, 1846.

Discover. "Google VR." Last updated January 11, 2019, https://developers.google.com /vr/discover/.

Dobraszczyk, Paul. "Sewers, Wood Engraving and the Sublime: Picturing London's Main Drainage System in the *Illustrated London News*, 1859–62." *Victorian Periodicals Review* 38, no. 4 (Winter 2005): 349–378.

Doran, Robert. *The Theory of the Sublime from Longinus to Kant*. Cambridge: Cambridge University Press, 2015.

Dorsen, Annie. "The Sublime and the Digital Landscape." *Theater* 48, no. 1 (2018): 55–67.

Dreiser, Theodore. *A Hoosier Holiday.* New York: Oxford University Press, 1932.

Dubois, W. E. B. *Darkwater: Voices from Within the Veil.* New York: Harcourt, Brace and Company, 1920.

Duffy, Cian. *The Landscapes of the Sublime, 1700–1830.* London: Palgrave MacMillan, 2013.

Dunaway, Finis. "Seeing Global Warming: Contemporary Art and the Fate of the Planet." *Environmental History* 14 (2009): 9–31.

Dunaway, Finis. *Seeing Green: The Use and Abuse of American Environmental Images.* Chicago: University of Chicago Press, 2015.

Du Preez, Amanda. "Sublime Selfies: To Witness Death." *European Journal of Cultural Studies* 21, no. 6 (2018): 744–760.

Durkheim, Émile. "The Elementary Forms of Religious Life." In *Durkheim on Religion,* ed. W. S. G. Pickering, 102–166. London: Routledge & Kegan Paul, 1975.

Economides, Louise. *The Ecology of Wonder in Romantic and Postmodern Literature.* London: Palgrave Macmillan, 2016.

Ellis, S. R. "Nature and Origins of Virtual Environments." *Computing Systems in Engineering* 2, no. 4 (1991): 321–327.

Ellison, Joshua, and Geoff Monaugh. "Simon Norfolk in Interview." In "The Media," *Irish Pages* 4, no. 1 (2007): 201–216.

Emerson, Ralph Waldo. "Country Life." In *The Later Lectures of Ralph Waldo Emerson, Volume 2: 1855–1891,* ed. Ronald A. Bosco and Joel Myerson, 49–67. Athens: University of Georgia Press, 2010.

Emmett, Robert S., and David E. Nye. *The Environmental Humanities: A Critical Introduction.* Cambridge, MA: MIT Press, 2017.

Evernden, Neil. *The Social Creation of Nature.* Baltimore: Johns Hopkins University Press, 1992.

Ferguson, Harvie. "The Sublime and the Subliminal: Modern Identities and the Aesthetics of Combat." *Theory, Culture, and Society* 21, no. 3 (2004): 1–33.

Finn, Megan. *Documenting Aftermath: Information Infrastructures in the Wake of Disasters.* Cambridge, MA: MIT Press, 2018.

Fisher, S. S., E. M. Wensel, C. Coler, and M. W. McGreevy. "Virtual Interface Environment Workstations." *Proceedings of the Human Factors Society* 32 no. 2 (1988). https://www.researchgate.net/publication/4709345_Virtual_Interface_Environment_Workstations.

Fletcher, Seth. "How Do You Take a Picture of a Black Hole? With a Telescope Big as the Earth." *New York Times*, October 4, 2018.

Frana, Philip L. Interview with Carl Machover for the Charles Babbage Institute, June 20, 2002. University of Minnesota Digital Conservancy. http://hdl.handle .net/11299/107462.

Friedman, Ted. *Electric Dreams*. New York: NYU Press, 2005.

Frierson, Patrick. "Introduction." In *Immanuel Kant, Observations on the Feeling of the Beautiful and Sublime and Other Writings*, vii–xxxv. Cambridge: Cambridge University Press, 2011.

Furlan, Rod. "See the Future through Microsoft's HoloLens Augmented-Reality Glasses." *IEEE Spectrum* 53, no. 6 (June 2016): 21–22.

Fyfe, Paul. "Illustrating the Accident: Railways and the Catastrophic Picturesque in the *London Illustrated News*." *Victorian Periodicals Review* 46, no. 1 (Spring 2013): 61–91.

Galant, Debra. "Driven to Distraction." *New York Times*, July 20, 2003.

Gallagher, Carole. *American Ground Zero: The Secret Nuclear War*. Cambridge, MA: MIT Press, 1993.

Garrard, Greg. "Memories of Snow: Nostalgia, Amnesia, Re-reading." In *Memory in the Twenty-First Century*, ed. Sebastian Gross, 163–169. London: Palgrave Macmillan, 2016.

Genthe, Arnold. "Sacramento Street, San Francisco, 1906." Photograph. Steinbrugge Collection, UC Berkeley Earthquake Engineering Research Center.

Gibson, William. *Burning Chrome*. London: HarperCollins, 2000.

Gibson, William. *Neuromancer*. New York: Grafton, 1986.

Gilkey, Charles W. "The Spiritual Uplift of Scenery as Exemplified by the Grand Canyon." In *Proceedings of the National Parks Conference*, 138–140. Washington, DC: Government Printing Office, 1917.

Goatcher, J., and V. Brunsden. "Chernobyl and the Sublime Tourist." *Tourist Studies* 11, no. 2 (2011): 115–137.

Gold, Sylviane. "Same New York Rivers, but All Else Was New, A Review of 'Industrial Sublime' at the Hudson River Museum in Yonkers." *New York Times*, November 1, 2013.

Goldthwait, John. "Translator's Introduction." In Immanuel Kant, *Observations of the Feeling of the Beautiful and Sublime*, 1–38. Berkeley: University of California Press, 1960.

Golinski, J. V. "Joseph Priestley and the Chemical Sublime in British Public Science." In *Science and Spectacle in the European Enlightenment*, ed. Bernadette Bensaude-Vincent and Christine Blondel, 117–128. Aldershot: Ashgate, 2008.

Gordon, Lewis. "Can Virtual Nature Be a Good Substitute for the Great Outdoors? The Science Says Yes." *Washington Post*, April 29, 2020.

Gowans, James Russell. "WSMR [White Sands Missile Range]." MA thesis, University of California, Irvine, 2017.

Gray, Chris Hables. "Drones, War, and Technological Seduction." *Technology and Culture* 59, no. 4 (2018): 954–962.

Gray, J. Glenn. *The Warriors: Reflections on Men in Battle*. Lincoln: University of Nebraska Press, 1970.

The Great Chicago Fire of 1871: Three Illustrated Accounts from Harper's Weekly. Ashland, OR: Lewis Osborne, 1969.

Greengard, Samuel. *Virtual Reality*. Cambridge, MA: MIT Press, 2019.

Gresham, John D. "Gulf War: The Battle of 73 Easting and the Road to the Synthetic Battlefield." *DefenseMediaNetwork*, February 22, 2011. https://www.defensemedia network.com/stories/gulf-war-20th-the-battle-of-73-easting-and-the-road-to-the -synthetic-battlefield/2/.

Greshko, Michael. "The Mars Rover Opportunity Is Dead." *National Geographic*, February 13, 2019.

Gross, Alan G. *The Scientific Sublime: Popular Science Unravels the Mysteries of the Universe*. Oxford: Oxford University Press, 2018.

Hales, Peter Bacon. "The Atomic Sublime." *American Studies* 32, no. 1 (Spring 1991): 5–31.

Hall, Roger, Gordon Dodds, Stanley Triggs, and William Notman. *The World of William Notman: The Nineteenth Century through a Master Lens*. Boston: David R. Godine, 1993.

Haraway, Donna. "Cyborgs to Companion Species." In *The Haraway Reader*, 295–320. London: Routledge, 2004.

Haraway, Donna. "A Manifesto for Cyborgs." In *The Haraway Reader*, 7–46. London: Routledge, 2004.

Haraway, Donna. "Morphing in the Order." In *The Haraway Reader*, 199–222. London: Routledge, 2004.

Hartman, Geoffrey. *Wordsworth's Poetry*. New Haven: Yale University Press, 1971.

Haselstein, Ulla. "Seen from a Distance: Moments of Negativity in the American Sublime." *Amerikastudien/ American Studies* 43, no. 3 (1998): 405–421.

Hata, Hiroki, Katsutoshi Watanabe, and Makoto Kato. "Geographic Variation in the Damselfish–Red Alga Cultivation Mutualism in the Indo-West Pacific." *BMC Evolutionary Biology* 10, 185 (2010).

Haw, Richard. *The Brooklyn Bridge: A Cultural History*. New Brunswick: Rutgers University Press, 2008.

Heath, Malcolm. "Longinus and the Ancient Sublime." In *The Sublime*, ed. Timothy M. Costelloe, 11–23. Cambridge: Cambridge University Press, 2012.

Hecht, Gabriella. *The Radiance of France: Nuclear Power and National Identity after World War II*. Cambridge, MA: MIT Press, 1998.

Heingman, Noah. "The Style of Natural Catastrophes." *Huntington Library Quarterly* 66, no. 1/2 (2003): 97–133.

Helmreich, Stefan. "Intimate Sensing." In *Simulation and Its Discontents*, ed. Sherry Turkle, 129–149. Cambridge, MA: MIT Press, 2009.

Hemez, Colem. "Photographic Visions of the Atomic Sublime." *Bowdoin Journal of Art* 4 (2018): 1–41.

Hemingway, Ernest. "Soldier's Home." In *In Our Time*, 87–101. New York: Charles Scribner's, 1925.

Henshaw, John M. *A Tour of the Senses*. Baltimore: Johns Hopkins University Press, 2012.

Hepburn, Ronald W. "Landscape and the Metaphysical Imagination." *Environmental Values* 5, no. 3 (August 1996): 191–204.

Higgins, Andrew. "No Bed, No Breakfast, but 4-Star Gunfire. Welcome to a War Hostel." *New York Times*, November 28, 2018.

Hitt, Christopher. "Toward an Ecological Sublime." *New Literary History* 30, no. 3 (1999): 603–623.

Hobsbawm, Eric, and Terence Ranger. *The Invention of Tradition*. Cambridge: Cambridge University Press, 1983.

Hoffman, Susanna M., and Anthony Oliver-Smith, eds. *Culture and Catastrophe: The Anthropology of Disaster*. Santa Fe, NM: The School of American Research Press, 2002.

Holman, Brett. "The Meaning of Hendon: The Royal Air Force Display, Aerial Theatre and the Technological Sublime, 1920–1937." *Historical Research* 93, no. 259 (February 2020): 131–152.

Horner, Victoria. "The Cultural Mind of Chimpanzees." In *The Mind of the Chimpanzee*, ed. Elizabeth V. Lonsdorf, Stephen R. Ross, and Tetsuro Matsuzawa, 101–115. Chicago: University of Chicago Press, 2010.

Howlett, Debbie. "Americans Driving to Distraction as Multitasking on Road Rises." *USA Today*, March 5, 2004.

Howley, Kevin. *Drones: Media Discourse and the Public Imagination*. New York: Peter Lang, 2018.

Huntemann, Nina B., and Matthew Thomas Payne. *Joystick Soldiers*. London: Routledge, 2010.

Jackson, J. B. "The Abstract World of the Hot-Rodder." In *Changing Rural Landscapes*, ed. Ervin H. and Margaret J. Zube, 140–151. Amherst: University of Massachusetts Press, 1977.

Jacobs, Emma. "On the Frontline of War-Zone Tourism." *Financial Times*, October 31, 2013.

Jager, Jens. "Picturing Nations: Landscape Photography and National Identity in Britain and Germany in the Mid-Nineteenth Century." In *Picturing Place*, ed. Joan M. Schwartz and James R. Ryan, 117–140. London: I. B. Taurus, 2003.

Johnson, Stephen. "Natural Magic." *New York Times Magazine*, November 6, 2016, 50–53.

Johnston, Sean. *Holograms: A Cultural History*. Oxford: Oxford University Press, 2015.

Johnston, Sean. *Holographic Visions: A History of New Science*. Oxford: Oxford University Press, 2016.

Jörg, Friedrich, and Allison Brown. *The Fire: The Bombing of Germany, 1940–1945*. New York: Columbia University Press, 2008.

Jørgensen, Dolly. *Recovering Lost Species in the Modern Age: Histories of Longing and Belonging*. Cambridge, MA: MIT Press, 2019.

Jørgensen, Dolly. "Rethinking Rewilding." *Geoforum* 65 (October 2015): 482–488. https://www.sciencedirect.com/science/article/abs/pii/S0016718514002504.

Junger, Sebastian. *War*. London: Fourth Estate, 2010.

Kamin, Debra. "The Rise of Dark Tourism." *The Atlantic*, July 15, 2014.

Kang, Minsoo. *Sublime Dreams of Living Machines*. Cambridge, MA: Harvard University Press, 2011.

Kant, Immanuel. *Critique of the Power of Judgment*, ed. Paul Guyer, trans. Paul Guyer and Eric Matthews. The Cambridge Edition of the Works of Immanuel Kant. Cambridge: Cambridge University Press, 2000.

Kant, Immanuel. *Observations of the Feeling of the Beautiful and Sublime*. Berkeley: University of California Press, 1960.

Kant, Immanuel. *Universal Natural History of the Heavens*. Arlington, Virginia: Richer Resources Publications, 2009.

Kaplan, Caren. *Aerial Aftermaths: Wartime from Above*. Durham: Duke University Press, 2018.

Kasson, John. *Civilizing the Machine*. Harmondsworth: Penguin, 1977.

Keegan, John. *The Face of Battle*. London: The Bodley Head, 2014.

Keltner, Dacher, et al. "Interpersonal Relations and Group Processes: Awe and Humility." *Journal of Personality and Social Psychology* 114, no. 2 (2018): 258–269.

Keltner, Dacher, and J. Haidt. "Approaching Awe, a Moral, Spiritual, and Aesthetic Emotion." *Cognition and Emotion* 17 (2003): 297–314.

Kemeny, Andras, Paul George, Frédéric Mérienne, and Florent Colombet. "New VR Navigation Techniques to Reduce Cybersickness." *Electronic Imaging* 3 (2017): 48–53.

Kessler, Elizabeth. *Picturing the Cosmos: Hubble Space Telescope Images and the Astronomical Sublime*. Minneapolis: University of Minnesota Press, 2012.

Kimmerer, Robin Wall. *Braiding Sweetgrass*. Minneapolis: Milkweed, 2013.

King, Brad, and John Borland. *Dungeons and Dreamers: The Rise of Computer Game Culture*. New York: McGraw Hill, 2003.

Kirshenbaum, Matthew. "Krigsspiel." In *Debugging Game History: A Critical Lexicon*, ed. Henry Lowood and Raiford Guins, 279–286. Cambridge, MA: MIT Press, 2016.

Klauber, Laurence Monroe. "Two Days in San Francisco, 1906." Witness statement, May 1, 1906. Online Archive of California. Bancroft Library, University of California, Berkeley. https://oac.cdlib.org/view?docId=hb1p3004c0&brand=oac4&doc.view =entire_text.

Kolbert, Elizabeth. *The Sixth Extinction*. New York: Henry Holt, 2015.

Kovacs, Claire. "Pompeii and Its Material Reproductions: The Rise of a Tourist Site in the Nineteenth Century." *Journal of Tourism History* 5, no. 1 (2013): 25–49.

Krapp, Peter. "Control." In *Debugging Game History: A Critical Lexicon*, ed. Henry Lowood and Raiford Guins, 73–80. Cambridge, MA: MIT Press, 2016.

Krapp, Peter. "Nomads of the Technical Sublime." In Marion Näser-Lather and Christoph Neubert, eds. *Traffic: Media as Infrastructures and Cultural Practices*, 205–219. Leiden: Brill, 2015.

Kupensky, Nicholas Kyle. "The Soviet Industrial Sublime: The Awe and Fear of Dneprostroi, 1927–1932." Ph.D. dissertation, Yale University, 2017.

Landau, S., and Carl Condit. *The Rise of the New York Skyscraper*. New Haven: Yale University Press, 1996.

Langford, Barry. "Seeing Only Corpses: Vision and/of Urban Disaster in Apocalyptic Cinema." In *Urban Space and Cityscapes*, ed. Christoph Lindner, 38–48. New York: Routledge, 2006.

Lanier, Jaron. *The Dawn of the New Everything*. London: Penguin Random House, 2017.

Lanier, Jaron. *You Are Not a Gadget*. New York: Knopf, 2010.

Larsen, Julie. "The Living Archive and the Sublime Nature of the Anthropocene: An Architecture Design Studio." *UPLand: Journal of Urban Planning, Landscape and Environmental Design* 2, no. 2 (2017): 223–238.

LeCain, Timothy J. *Mass Destruction: The Men and Giant Mines That Wired America and Scarred the Planet*. New Brunswick: Rutgers University Press, 2009.

Leech, Brian. "Protest, Power, and the Pit: Fighting Open-Pit Mining in Butte, Montana." *Montana: The Magazine of Western History* 62, no. 2 (Summer 2012): 24–43, 90–94.

Leiner, Barry M., Vinton G. Cerf, David D. Clark, Robert E. Kahn, Leonard Kleinrock, Daniel C. Lynch, Jon Postel, Larry G. Roberts, and Stephen Wolff. "Brief History of the Internet." Internet Society, 1997. https://www.internetsociety.org/internet/history-internet/brief-history-internet/.

Leopold, Aldo. "The Round River." In *A Sand County Almanac, with Essays on Conservation from Round River*. New York: Ballantine, 1970.

Licht, Melissa Vera. "Warring Opinions: An Investigation into the Sublime Aesthetic Narratives of Contemporary Warfare." PhD dissertation, University of Minnesota, 2010.

Lifton, Robert Jay. *Witness to an Extreme Century*. New York: Free Press, 2011.

Lilienthal, Edward Tabor. *Sacred Ground: Americans and Their Battlefields*. Urbana: University of Illinois, 1993.

Linzey, Andrew, ed. *The Global Guide to Animal Protection*. Urbana: University of Illinois, 2013.

Littmann, William. "The Production of Goodwill: The Origins and Development of the Factory Tour in America." *Perspectives in Vernacular Architecture* 9 (2003): 71–84.

Lopez, Barry. *Arctic Dreams*. New York: Charles Scribner's, 1986.

Lopez, Barry, ed. *The Future of Nature*. Minneapolis: Milkweed Editions, 2007.

Lopez, Barry. *Horizon*. New York: Alfred A Knopf, 2019.

Lowood, Henry, and Raiford Guins. *Debugging Game History: A Critical Lexicon*. Cambridge, MA: MIT Press, 2016.

Lyotard, Jean-François. "The Sublime and the Avantgarde." In *The Inhuman*, 89–107. Cambridge: Polity Press, 1991.

Macfarlane, Robert. *Underland*. London: Penguin/Random House, 2019.

MacRumors Staff. "Apple Glasses." *MacRumors*, September 22, 2021. https://www.macrumors.com/roundup/apple-glasses/.

Malmstrom, Carolyn M. "Ecologists Study the Interactions of Organisms and Their Environment." *Nature Education Knowledge Project* 3, no. 10 (2010): 88. https://

www.nature.com/scitable/knowledge/library/ecologists-study-the-interactions-of
-organisms-and-13235586/.

Manback, John B. *Brooklyn: Historically Speaking*. Mt. Pleasant, SC: Arcadia Publishing, 2008.

Marchand, Roland. *Advertising the American Dream*. Berkeley: University of California Press, 1985.

Marlowe, Emily. "Natural and Technological Wonders: Embracing Modernity at Carlsbad Caverns National Park." PhD dissertation, Drew University, 2016.

Marx, Leo. *The Machine in the Garden*. New York: Oxford University Press, 1965.

Mauch, Christof, and Thomas Zeller, eds. *The World Beyond the Windshield: Roads and Landscapes in the United States and Europe*. Columbus: Ohio University Press, 2008.

Mayor, Andrienne. *Gods and Robots: Myths, Machines and Ancient Dreams of Technology*. Princeton: Princeton University Press, 2018.

McCarthy, Cormac. *The Road*. London: Picador, 2006.

McClelland, Linda Flint. *Building the National Parks*. Baltimore: Johns Hopkins University Press, 1998.

McCullough, David. *The Johnstown Flood*. New York: Simon and Schuster, 1968.

McHaney, Roger. "Development of a Generic Simnet II Simulation Package for Unidirectional, Sideloading AGV Systems." In *Proceedings of the 1995 Winter Simulation Conference*, ed. C. Alexopoulos, K. Kang, W. R. Lilegdon, and D. Goldsman, 823–828. Piscataway, NJ: IEEE, 1995.

McKibben, Bill. "Designer Genes." In *The Future of Nature*, ed. Barry Lopez, 257–267. Minneapolis: Milkweed Editions, 2007.

McKinsey, Elizabeth. *Niagara Falls: Icon of the American Sublime*. Cambridge: Cambridge University Press, 1985.

Miller, Duncan C., and Jack A. Thorpe. "SIMNET: The Advent of Simulator Networking." *Proceedings of the IEEE* 83, no. 8 (August 1995): 1114–1123.

Miller, Perry. *The Life of the Mind in America: From the Revolution to the Civil War*. San Diego: Harcourt Brace Jovanovich, 1965.

Miller, Shawn William. *An Environmental History of Latin America*. Cambridge: Cambridge University Press, 2007.

Missal, Alexander. *Seaway to the Future: American Social Visions and the Construction of the Panama Canal*. Madison: University of Wisconsin Press, 2008.

Monk, Samuel Holt. *The Sublime: A Study of Critical Theories in Eighteenth Century England*. Ann Arbor: University of Michigan Press, 1960.

Morshed, Adnan. *Impossible Heights: Skyscrapers, Flight and the Master Builder*. Minneapolis: University of Minnesota, 2015.

Morton, Timothy. *Being Ecological*. London: Penguin Books, 2018.

Morton, Timothy. *Ecology without Nature: Rethinking Environmental Aesthetics*. Cambridge, MA: Harvard University Press, 2007.

Mosco, Vincent. *The Digital Sublime*. Cambridge, MA: MIT Press, 2004.

Munk, Lee Ann, Scott A. Hynek, Dwight C. Bradley, David Boutt, Keith Labay, and Hillary Jochens. "Lithium Brines: A Global Perspective." In *Rare Earth and Critical Elements in Ore Deposits*, ed. Philip L. Verplanck and Murray W. Hitzman, 339–346. Reviews in Economic Geology, vol. 18. Littleton, CO: Society of Economic Geologists, 2016.

Myers, David. "Simulation." In *Debugging Game History: A Critical Lexicon*, ed. Henry Lowood and Raiford Guins, 293–400. Cambridge, MA: MIT Press, 2016.

Mynott, Sara. "Newly Discovered Hermit Crab Species Lives in 'Walking Corals.'" *The Conversation*. September 20, 2017. https://theconversation.com/newly-discovered -hermit-crab-species-lives-in-walking-corals-84389.

Natarajan, Priyamvada. "At Long Last, a Glimpse of a Black Hole." *New York Times*, April 8, 2019.

Nayar, Pramod K. "The Imperial Sublime: English Travel Writing and India, 1750–1920." *Journal of Early Modern Cultural Studies* 2, no. 2 (2002): 55–99.

Negroponte, Nicholas. *Being Digital*. New York: Vintage, 1995.

Niccol, Andrew, dir. *Good Kill*. Voltage Pictures and Sobrini Films, 2014. 102 min.

Nichols, Randy. "Target Acquired: America's Army and the Video Games Industry." In *Joystick Soldiers*, ed. Nina B. Huntemann and Matthew Thomas Payne, 39–52. London: Routledge, 2010.

Nixon, Rob. *Slow Violence and the Environmentalism of the Poor*. Cambridge, MA: Harvard University Press, 2011.

Nurmis, Joanna. "Between Dread and Delight: Motivational Discourses in Climate Change." Presented at *Bridging Divides: Spaces of Scholarship and Practice in Environmental Communication*, Conference on Communication and Environment. Boulder, Colorado, June 11–14, 2015. https://theieca.org/coce2015.

Nurmis, Joanna. "Visual Climate Change Art 2005–2015: Discourse and Practice." *WIREs Climate Change* 7 (2016): 501–516.

Nye, David E. *America's Assembly Line*. Cambridge, MA: MIT Press, 2013.

Nye, David E. *American Illuminations: Urban Lighting, 1800–1920*. Cambridge, MA: MIT Press, 2018.

Nye, David E. *American Technological Sublime*. Cambridge, MA: MIT Press, 1994.

Nye, David E. *Conflicted American Landscapes*. Cambridge, MA: MIT Press, 2021.

Nye, David E. *Electrifying America: Social Meanings of a New Technology*. Cambridge, MA: MIT Press, 1990.

Nye, David E. *Image Worlds: Corporate Identities at General Electric*. Cambridge, MA: The MIT Press, 1985.

Nye, David E. "Superfund Sites as Anti-landscapes." In *Contesting Environmental Imaginaries: Nature and Counternature in a Time of Global Change*, ed. Steve Hartman, 283–305. Leiden: Brill, 2017.

Nye, David E. "What Comes after the Technological Sublime?" In "Technology and the Sublime," ed. Giulia Rispoli and Christoph Rosol, *Azimuth* 6, no. 12 (2018): 167–183.

Nye, David E. *When the Lights Went Out*. Cambridge, MA: MIT Press, 2010.

Nye, David E., and Sarah Elkind, eds. *Antilandscapes*. Amsterdam: Rodopi/Brill, 2014.

O'Brien, Tim. "How to Tell a True War Story." In *The Things They Carried*, 84–91. Boston: Houghton Mifflin, 1990.

O'Connell, Mark. "Journey to the End of the World." *New York Times Magazine*, March 29, 2020, 34-43.

Oettermann, Stephan. *The Panorama: History of a Mass Medium*. Cambridge, MA: Zone Books, 1997.

Oppermann, Serpil, and Serenella Iovino. "The Environmental Humanities and the Challenges of the Anthropocene." In *Environmental Humanities: Voices from the Anthropocene*, ed. Serpil Oppermann and Serenella Iovino, 21–32. Lanham, MD: Rowman & Littlefield, 2016.

Orvell, Miles. "After 9/11: Photography, the Destructive Sublime, and the Postmodern Archive." *Michigan Quarterly Review* 45, no. 2 (Spring 2006): 239–256.

Orvell, Miles. *American Photographs*. Oxford: Oxford University Press, 2003.

Orvell, Miles. "Cultural Studies, Visual Studies, Historical Studies: Connecting the Dots." *American Studies* 47, no. 2 (Summer 2006): 87–103.

Orvell, Miles. *Empire of Ruins: American Culture, Photography, and the Spectacle of Destruction*. New York: Oxford University Press, 2021.

Orvell, Miles. "Photographing Disaster: Urban Ruins and the Destructive Sublime." *Amerikastudien/American Studies* 58, no. 4 (December 2013): 647–671.

Orvell, Miles. *Photography in America*. New York: Oxford University Press, 2016.

Orvell, Miles. "Virtual Culture and the Logic of American Technology." *Revue française d'études américaines* 76 (March 1998): 12–27.

Overbye, Dennis. "Darkness Visible, Finally: Astronomers Capture First Ever Image of a Black Hole." *New York Times*, April 10, 2019.

Overgaard, Sidsel. "To Keep African Swine Fever Out, Denmark Is Planning a Southern Boar(der) Fence." National Public Radio, January 27, 2019. https://www.npr .org/2019/01/27/688152778/to-keep-african-swine-fever-out-denmark-is-planning-a -southern-boar-der-fence?t=1579874247249.

Owen, David. *Green Metropolis*. New York: Riverhead Books, 2009.

Pankhuri, Yadav. "Train Crushes 2 Boys Clicking Selfies in Delhi." January 17, 2017. https://timesofindia.indiatimes.com/city/delhi/train-crushes-2-boys-clicking-selfies -in-delhi/articleshow/56606607.cms.

Parker, Martin. "Vertical Capitalism: Skyscrapers and Organization." *Culture and Organization* 21, no. 3 (2015): 217–234.

Pavlovskis, Zoja. *Man in an Artificial Landscape: The Marvels of Civilization in Imperial Roman Literature*. Leiden: Mnemosyne Biblioteca Classica Catava, 1973.

Peebles, Jennifer. "Toxic Sublime: Imagining Contaminated Landscapes." *Environmental Communication* 5, no. 4 (December 2011): 373–392.

Peña, Nonny de la, Peggy Weil, Joan Llobera, and Elias Giannopoulos. "Immersive Journalism: Immersive Virtual Reality for the First-Person Experience of News." *Presence* 19, no. 4 (2010): 291–301.

Pennington, Emily. "Here to Help, Amp Up Your Awe with a Microadventure." *New York Times*, July 22, 2021, A3.

Peterson, Dale. *Jane Goodall: The Woman Who Redefined Man*. Boston: Houghton Mifflin, 2006.

Pezzullo, Phaedra C. "'This Is the Only Tour That Sells:' Tourism, Disaster, and National Identity in New Orleans." *Journal of Tourism and Cultural Change* 7, no. 2 (2009): 99–114.

Price, Derrick. *Coal Cultures: Picturing Mining Landscapes and Communities*. London: Bloomsbury, 2018.

Pringle, Robert M., Daniel F. Doak, Alison K. Brody, Rudy Jocqué, and Todd M. Palmer. "Spatial Pattern Enhances Ecosystem Functioning in an African Savanna." *PLoS Biol* 8 (May 2010). https://www.ncbi.nlm.nih.gov/pmc/articles/PMC2876046/.

Rabinovitz, Lauren. *Electric Dreamland: Amusement Parks, Movies, and American Modernity*. New York: Columbia University Press. 2012.

Reingold, Howard. *The Virtual Community*. Cambridge, MA: MIT Press, 2000.

Renan, Ernest. "What Is a Nation?" Lecture at the Sorbonne, March 11, 1882. Republished in Ernest Renan, *Qu'est-ce qu'une nation?* Paris: Presses-Pocket, 1992.

Ripley, Amanda. *The Unthinkable: Who Survives When Disaster Strikes—And Why*. New York: Three Rivers Press, 2009.

Rodgers, Anastasia. "Constructing Beauty: The Photographs Documenting the Construction of the Bloor Viaduct." *Archivaria* 54 (2002): 72–91.

Romains, Jules. *Verdun*, trans. Gerard Hopkins. London: Prion, 2000.

Rosenfield, Michael. "We've Reached the Summit." *IBM Research Blog*, June 14, 2018. https://www.ibm.com/blogs/research/2018/06/summit/.

Roszak, Theodore. *The Cult of Information*. Berkeley: University of California Press, 1994.

Rozario, Kevin. *The Culture of Calamity*. Chicago: University of Chicago Press, 2007.

Rozelle, Lee. *Ecosublime: Environmental Awe and Terror from New World to Oddworld*. Tuscaloosa: University of Alabama Press, 2006.

Ruiz, Susanne. "Darfur Is Dying." Accessed September 3, 2021. https://susanaruiz.org/takeactiongames-darfurisdying.

Ryan, Marie-Laure. "Narrative." In *Debugging Game History: A Critical Lexicon*, ed. Henry Lowood and Raiford Guins, 335–342. Cambridge, MA: MIT Press, 2016.

Sayre, Gordon M. "If Jefferson Had Visited Niagara Falls: The Sublime Wilderness Spectacle in America, 1775–1825." *Interdisciplinary Studies in Literature and Environment* 8, no. 2 (Summer 2001): 141–162.

Schechner, Richard. "9/11 as Avant-Garde Art?" *PMLA* 124, no. 5 (October 2009): 1820–1829.

Scranton, Roy. *Learning to Die in the Anthropocene: Reflections on the End of Civilization*. San Francisco: City Lights, 2015.

Sears, John. *Sacred Places: American Tourist Attractions in the Nineteenth Century*. New York: Oxford University Press, 1989.

"Selfie Deaths: 259 People Reported Dead Seeking the Perfect Picture." BBC News, October 4, 2018. https://www.bbc.com/news/newsbeat-45745982.

Shapshay, Sandra. "Contemporary Environmental Aesthetics and the Neglect of the Sublime." *British Journal of Aesthetics* 53, no. 2 (April 2013): 181–198.

Shaw, Ian G. R. *Predator Empire: Drone Warfare and Full Spectrum Dominance*. Minneapolis: University of Minnesota Press, 2016.

Sherry, Michael S. *The Rise of American Air Power*. New Haven: Yale University Press, 1987.

Shinkle, Eugénie. "Video Games and the Digital Sublime." In *Digital Cultures and the Politics of Emotion*, ed. Athina Karatzogianni and Adi Kuntsman, 94–108. London: Palgrave Macmillan, 2012.

Shusterman, Richard. "Somaesthetics and Burke's Sublime." *British Journal of Aesthetics* 45, no. 4 (October 2005): 323–341.

Slater, M., D. Perez-Marcos, H. Ehrsson, and M. V. Sanchez-Vives. "Inducing Illusory Ownership of a Virtual Body." *Frontiers in Neuroscience* 3, no. 2 (2009): 214–220.

Sledge, E. B. *With the Old Breed at Peleliu and Okinawa.* Novato, CA: Presideo Press, 1981.

Sluka, Jeffrey A. "Drones in the Tribal Zone." In *War, Technology, Anthropology*, ed. Koen Stroeken, chapter 1. Oxford: Berghahn Books, 2012.

Smith, Carl. "Faith and Doubt: The Imaginative Dimensions of the Great Chicago Fire." In *American Disasters*, ed. Steven Biel, 129–169. New York: New York University Press, 2001.

Smith, Saphora. "Marc Quinn's 'The Toxic Sublime' Shows a Provocateur and Social Observer." *New York Times*, August 16, 2015.

Smith, Sarah, Mark Ottoni-Wilhelm, and Kimberley Scharf. "The Donation Response to Natural Disasters." Institute for Fiscal Studies, Working Paper W17/19, 2017.

Solnit, Rebecca. *A Field Guide to Getting Lost.* Edinburgh: Canongate Books, 2005.

Sonter, Laura J., Diego Herrera, Damian J. Barrett, Gillian L. Galford, Chris J. Moran, and Britaldo S. Soares-Filho. "Mining Drives Extensive Deforestation in the Brazilian Amazon." *Nature Communications*, October 18, 2017. https://www.nature.com/articles/s41467-017-00557-w.

"South America's Drought-Hit Paraná River at 77-Year Low." BBC News, September 1, 2021. https://www.bbc.com/news/world-latin-america-58408791.

Spinola, Julia. "Monstrous Art." *Frankfurter Allgemeine Zeitung*, September 25, 2001.

Stein, Blair Rebecca. "All That Is Solid Melts into Air Travel: Environments, Technologies, and the Modern Nation at Trans-Canada Air Lines." PhD dissertation, University of Oklahoma, 2019.

Steinberg, Ted. *Acts of God: The Unnatural History of Natural Disaster in America.* New York: Oxford University Press, 2000.

Steller, Georg. *Steller's History of Kamchatka.* Fairbanks: University of Alaska Press, 2003.

Stilgoe, John. *Metropolitan Corridor: Railroads and the American Scene.* New Haven: Yale University Press, 1983.

Stone, Allucquère Rosanne. *The War of Desire and Technology.* Cambridge, MA: MIT Press, 1995.

Strand, Ginger. *Inventing Niagara.* New York: Simon & Schuster, 2008.

Styran, Roberta, and Robert R. Taylor. *This Colossal Project: Building the Welland Ship Canal, 1913–1932*. Montreal: McGill University Press, 2016.

Sutter, Paul S. *Driven Wild*. Seattle: University of Washington Press, 2002.

Suzuki, David, and Amanda McConnell. *The Sacred Balance: Rediscovering Our Place in Nature*. Vancouver: Graystone Books, 1999.

Tabbi, Joseph. *Postmodern Sublime: Technology and American Writing from Mailer to Cyberpunk*. Ithaca, NY: Cornell University Press, 1995.

Tandt, Christophe Den. *The Urban Sublime in American Literary Naturalism*. Chicago: University of Illinois Press, 1998.

Tarantino, Matteo. "Toward a New 'Electrical World': Is There a Chinese Technological Sublime?" In *New Connectivities in China: Virtual, Actual and Local Interactions*, 185–200. London: Springer, 2012.

Taylor, Nicholas. "The Awful Sublimity of the Victorian City." In *The Victorian City, Images and Realities*, vol. 2, ed. H. J. Dyos and M. Wolff, 431–447. London: Routledge Kegan Paul, 1973.

Thébaud-Sorger, Marie. Review of *The Sublime Invention: Ballooning in Europe, 1783–1820*, by Michael Lynn. *Technology and Culture* 53, no. 2 (April 2012): 488.

Turkle, Sherry. *Simulation and Its Discontents*. Cambridge, MA: MIT Press, 2010.

UNISDR (United Nations International Strategy for Disaster Reduction Secretariat). *Global Assessment Report on Disaster Risk Reduction*, 2009. https://www.unisdr.org/we /inform/publications/9413.

United States Fish & Wildlife Service. "Migratory Fish Restoration Benefits Everyone." Last updated December 19, 2018. https://www.fws.gov/northeast/cnefro /Species.html.

Valance, Hélene. "Destructive Re-creations: Spectacles of Urban Destruction in Turn-of-the-Century United States." In *Sensationalism and the Genealogy of Modernity*, ed. Alberto Gabriele, 121–141. New York: Palgrave MacMillan, 2017.

Vanberburg, Colin. "Drone Art." *Dissent* 63, no. 3 (Summer 2016): 6–10.

Vanian, Jonathan. "Amazon Takes a Trip in Augmented and Virtual Reality." *Fortune*, November 27, 2017. https://fortune.com/2017/11/27/amazon-virtual-reality -augmented-sumerian/.

Vardi, Itali. "Auto Thrill Shows and Destruction Derbies, 1922–1965: Establishing the Cultural Logic of the Deliberate Car Crash in America." *Journal of Social History* 45, no. 1 (Fall 2011): 20–46.

Vertesi, Janet. "Seeing Like a Rover: Visualization, Embodiment, and Interaction on the Mars Exploration Rover Mission." *Social Studies of Science* 42, no. 3 (2012): 393–414.

Wakefield, Stephanie. "Infrastructures of Liberal Life: From Modernity and Progress to Resilience and Ruins." *Geography Compass* 12, no. 7 (July 2018): 123–177. https://onlinelibrary.wiley.com/doi/full/10.1111/gec3.12377.

Wasson, Haidee. "The Networked Screen: Moving Images, Materiality, and the Aesthetics of Size." In *Fluid Screen, Expanded Cinema*, ed. J. Marchessault and S. Lords, 74–95. Toronto: University of Toronto Press, 2007.

Watson, John. *The Philosophy of Kant Explained*. Facsimile reprint of 1908 edition. New York: Garland, 1976.

Weiskel, Thomas. *The Romantic Sublime: Studies in the Structure and Psychology of Transcendence*. Baltimore: Johns Hopkins University Press, 1976.

White, Leanna, and Elspeth Frew. *Dark Tourism and Place Identity: Managing and Interpreting Dark Places*. London: Routledge, 2013.

Whittle, Patrick. "Ambitious New Plan to Save Atlantic Salmon Has a Big Price Tag." *Concord Monitor*, February 14, 2019.

Williams, Terry Tempest. *Red: Passion and Patience in the Desert*. New York: Vintage Books, 2002.

Willim, Robert. "Imperfect Imaginaries: Digitization, Mundanisation, and the Ungraspable." In *Digitization: Theories and Concepts for Empirical Cultural Research*, ed. Gertraud Koch, 53–77. London: Routledge, 2017.

Willis, John. "William Notman: Witness and Photographer of the Victorian Era." Canadian Museum of History, November 22, 2018. https://www.historymuseum.ca/blog/william-notman-witness-and-photographer-of-the-victorian-era/.

Wilson, Byron. "The Sublime Anthropocene." *Environmental Philosophy* 13, no. 2 (2016): 155–174.

Wilson, Rob. "The Postmodern Sublime: Local Definitions, Global Deformations of the US National Imaginary." *Amerikastudien / American Studies* 43, no. 3 (1998): 517–527.

Wilson, Rob. *American Sublime: The Genealogy of a Poetic Genre*. Madison: Wisconsin University Press, 1991.

Winkler, Allan. *Life Under a Cloud: American Anxiety About the Atom*. New York: Oxford University Press, 1993.

Wohl, Robert. *A Passion for Wings: Aviation and the Western Imagination*. New Haven: Yale University Press, 1994.

Wohlleben, Peter. *The Hidden Life of Trees: What They Feel, How They Communicate*. New York: Vintage Penguin, 2016.

Woo, Lillian C. "Ecomimesis: A Sustainable Design Paradigm. Ecomimetic Solutions to Ecosystem Homeostatic Imbalances." *European Journal of Sustainable Development* 6, no. 4 (2017): 255–266.

Wood, Graeme. "My Atomic Holiday." *Atlantic*, September 2012. https://www .theatlantic.com/magazine/archive/2012/09/my-atomic-holiday/309059/.

Wormbs, Nina. "Sublime Satellite Imagery as Environing Technology." In "Technology and the Sublime," ed. Giulia Rispoli and Christoph Rosol, *Azimuth* 6, no. 12 (2018): 77–92.

Worster, Donald. *Nature's Economy: A History of Ecological Ideas*. Cambridge: Cambridge University Press, 1994.

Wray, Matt. "A Blast from the Past: Preserving and Interpreting the Atomic Age." *American Quarterly* 58, no. 2 (June 2006): 467–483.

Yablon, Nick. *Untimely Ruins: An Archaeology of American Urban Modernity, 1819–1919*. Chicago: University of Chicago Press, 2009.

Yankovska, Ganna, and Kevin Hannam. "Dark and Toxic Tourism in the Chernobyl Exclusion Zone." *Current Issues in Tourism* 17, no. 10 (2014): 929–939.

Younger, Rina. *Industry in Art: Pittsburgh, 1812 to 1920*. Pittsburgh: University of Pittsburgh Press, 2011.

Yost, Jeffrey R. Interview with Donn Parker for the Charles Babbage Institute, May 2003. University of Minnesota Digital Conservancy.

Yu, Xinge, Zhaoqian Xie, Yang Yu, Jungyup Lee, Abraham Vazquez-Guardado, et al. "Skin-Integrated Wireless Haptic Interfaces for Virtual and Augmented Reality." *Nature* 575, November 21, 2019, 473–479.

Zimmer, Carl. "A New Company with a Wild Mission." *New York Times*, September. 13, 2021. https://www.nytimes.com/2021/09/13/science/colossal-woolly-mammoth -DNA.html.

Zimmerman, Robert. *The Universe in a Mirror: The Saga of the Hubble Space Telescope and the Visionaries Who Built It*. Princeton: Princeton University Press, 2008.

Zuboff, Shoshana. *The Age of Surveillance Capitalism*. New York: Public Affairs, 2019.

Zytaruk, George J., and James T. Boulton, eds. *The Letters of D. H. Lawrence*, vol. 2. Cambridge: Cambridge University Press, 1981.

INDEX